Design and Management of Energy-Efficient Hybrid Electrical Energy Storage Systems

Younghyun Kim • Naehyuck Chang

Design and Management of Energy-Efficient Hybrid Electrical Energy Storage Systems

Younghyun Kim
Purdue University
West Lafayette, IN, USA

Naehyuck Chang
Seoul, Korea
Republic of (South Korea)

ISBN 978-3-319-36022-5 ISBN 978-3-319-07281-4 (eBook)
DOI 10.1007/978-3-319-07281-4
Springer Cham Heidelberg New York Dordrecht London

Printed on acid-free paper

Springer is part of Springer Science+Business Media (www.springer.com)

This book is dedicated to my beloved wife Jungsun, for her love and trust; and my parents, for their devotion and support.

Younghyun Kim

Preface

Electrical energy storage (EES) systems provide various benefits of high energy efficiency, high reliability, low cost, and so on, by storing and retrieving energy on demand. EES systems have a wide range of applications, such as contingency service and peak shaving for power grid, energy buffer for renewable power sources, power train in electric vehicles (EVs), and so on. Current EES systems mainly rely on a single type of energy storage technology, but unfortunately no single type of EES element technology can fulfill all the desirable characteristics, such as high power/energy density, low cost, high cycle efficiency, and long cycle life. Hybrid EES (HEES) systems, on the other hand, are composed of multiple, heterogeneous EES element technologies, aiming at exploiting the strengths of each technology while hiding its weaknesses. This is a practical approach to improve the performance of EES systems with currently available EES element technologies. A HEES system may achieve a combination of performance metrics that is superior to those for any of its individual energy storage elements with elaborated system design and control schemes.

This book introduces high-level design and management techniques that maximize the energy efficiency of HEES systems. We propose new architectures for HEES systems and systematic design optimization methods. The proposed networked charge transfer interconnect (CTI) architecture and bank reconfiguration architecture minimize power conversion loss and thus maximize the charge transfer efficiency in HEES systems. We also point out the limitations of conventional control schemes and propose a joint optimization design and control technique that simultaneously considers power sources. The proposed maximum power transfer tracking (MPTT) operation and MPTT-aware design method effectively increases energy harvesting efficiency and actual available energy. We finally introduce a prototype of a HEES system implementation that physically proves the feasibility of the proposed HEES system.

The content of this book describes work that had been carried out during Dr. Younghyun Kim's Ph.D. study in the CAD-X Lab at Seoul National University, led by Prof. Naehyuck Chang. Various chapters or sections of this book are based on scientific papers published in various conference proceedings and journals. We would like to acknowledge the work of the co-authors of those papers. Especially, large parts of the work have been conducted in collaboration with SPORT Lab at University of Southern California, led by Prof. Massoud Pedram.

This research was sponsored by the National Research Foundation of Korea (NRF) grant funded by the Korean Government (MEST) (No. 2014023320). The Institute of Computer Technology at Seoul National University provided research facilities for this study.

West Lafayette, USA — Younghyun Kim
Seoul, South Korea — Naehyuck Chang

Contents

Acronyms

AC	Alternating current
CPU	Central processing unit
CRM	Current regulation mode
CSR	Current source region
CTI	Charge transfer interconnect
DC	Direct current
DoD	Depth of discharge
DRAM	Dynamic random access memory
EDA	Electronic design automation
EES	Electrical energy storages
EV	Electrical vehicle
FPGA	Field-programmable gate array
GBRA	General balanced reconfiguration architecture
HDD	Hard disk drive
HEES	Hybrid electrical energy storages
HEV	Hybrid electrical vehicle
MOSFET	Metal-oxide-semiconductor field-effect transistor
MPP	Maximum power point
MPPT	Maximum power point tracking
MPT	Maximum power transfer
MPTT	Maximum power transfer tracking
NiCd	Nickel-cadmium
NiMH	Nickel-metal hydride
NoC	Network-on-chip
PCB	Printed circuit board
PID	Proportional-integral-derivative
PV	Photovoltaic
PWM	Pulse width modulation
RMS	Root mean square

SoC	State of charge
SoH	State of health
SRAM	Static random access memory
VLSI	Very-large-scale integration
VRM	Voltage regulation mode
VSR	Voltage source region
PP	Payback period

Chapter 1
Introduction

Electrical energy is a high quality form of energy, which can be easily and efficiently converted into other forms of energy and used to control other lower quality forms of energy [6]. Transformity of energy, which is the amount of one type of energy required to produce a heat equivalent of another type of energy, is an important factor of the quality of energy. Energy with higher transformity requires a larger amount of sunlight for its production, and therefore is more economically useful [7]. Electrical energy can be efficiently transformed into other forms of energy, with a transformity of 1.59×10^5 seJ/J (solar emjoules per joule), which is 2–4 times higher than that of fossil fuel energy [1].

However, effective and efficient use of electrical energy is not an easy problem. For grid-connected large-scale power systems, it is critical to maintain the stability of the power system, which is achieved by precisely balancing the generation and demand of electricity. Energy demand continuously changes over time, and long-term (daily or monthly) as well as short-term (secondly and minutely) changes should be followed by proper adjustment of power generation. However, the level of power generation of fossil fuel power plants and nuclear power plants cannot be adjusted immediately, and so they are not capable of dealing with the rapid change in the demand. Recent deployment of renewable power sources, which are considered promising sustainable energy, makes the situation even worse. The level of power generation of the renewable power sources, such as solar cells and windmills, is heavily dependent on environmental factors (e.g., the solar irradiance level or climate conditions) that are not controllable.

Electrical energy storage (EES) systems are a promising solution to address such challenges by storing energy and supplying the energy when needed at a high response speed with high efficiency. An EES is able to mitigate the mismatch between the power generation and demand without actually increasing the power generation, and prevents over-investment in power generation facilities and energy waste. By employing EES systems we can significantly improve the availability, efficiency, and stability of the power supply in a cost-effective way.

Y. Kim, N. Chang, *Design and Management of Energy-Efficient Hybrid Electrical Energy Storage Systems*, DOI 10.1007/978-3-319-07281-4_1

Despite their potential benefits, however, EES systems are still short for wide deployment. So far, development of a better EES element technology (e.g., battery and supercapacitor technology) has been focused as the key to success of high-performance EES systems. However, despite active research on the new EES element technologies, it is not likely for us to have an ultimate high-efficiency, high-power/energy capacity, low-cost, light-weight, and long-cycle life EES elements in the near future. In other words, there is no single type of EES element that can fulfill all the desired requirements to build an ideal EES system, as of today. As a result, current EES systems that are composed of homogeneous EES elements are subject to the limitations of the EES elements that are used. For example, battery-based EES systems requires regular battery replacement after hundreds of charging and discharging cycles; supercapacitor-based EES systems are cost-prohibitive for a large-scale deployment.

It is therefore practical to develop a system-level design methodology that enhances the EES system performance through efficient use of diverse energy storage technologies that are currently available. *Hybrid electrical energy storage (HEES) systems* consist of multiple, heterogeneous EES elements in order to exploit the advantages of the EES elements while hiding their shortcomings with aid of system-level managements [3, 9]. This approach aims to achieve the combination of performance metrics that are superior to that for any of its individual EES components from the heterogeneity. However, such benefits can be achieved only by elaborate optimizations during design and operation. Since a HEES system is composed of heterogeneous EES elements, the complexity of design and operation is much higher than that of conventional homogeneous EES systems, and its optimization is not trivial.

Interestingly, optimization problems of HEES systems have notable analogies with the memory subsystem of computers. Table 1.1 shows some representative computer memory devices. SRAM has the lowest latency and highest throughput, but is expensive and has a low capacity. On the other hand, DRAM is inferior to SRAM in terms of latency and throughput, but is cheaper and has a higher capacity. Mass storage devices such as HDD and flash memory have an even lower cost, higher capacity, non-volatility, but are subject to limited random access capability and write count. Composing the required memory space with SRAM only is infeasible due to its high cost except for supercomputers where cost is

Table 1.1 Comparison of memory devices [8]

		Cost	Current (mA)		Random access time (16 bit)			
	Density	($/Gb)	Idle	Active	Read	Write	Erase	Non-volatile
Fast SRAM	Low	614	5	65	10 ns	10 ns	10 ns	No
Low power SRAM	Low	320	0.005	3	55 ns	55 ns	55 ns	No
Mobile synch. DRAM	High	48	0.5	75	90 ns	90 ns	90 ns	No
NOR flash	High	96	0.03	32	200 ns	210.5 μs	1.2 s	Yes
NAND flash	Very high	21	0.01	10	10.1 μs	200.5 μs	2 ms	Yes

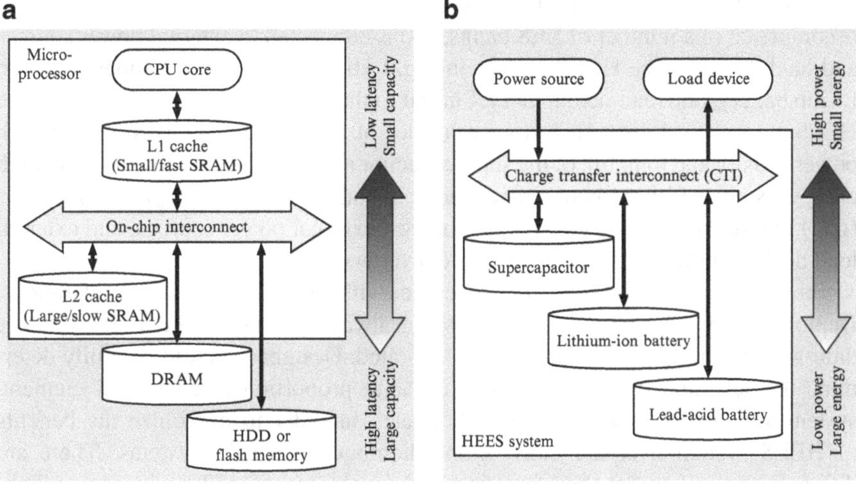

Fig. 1.1 Comparison of computer memory hierarchy and HEES system architecture. (**a**) Computer memory hierarchy. (**b**) HEES system

not a primary issue. On the other hand, using HDD or flash memory only cannot meet latency and throughput requirements of the CPU core and suffers from poor random access capability and limited write count. Computer architects, therefore, have remedied this problem by building a hierarchy of different types of memory devices. A typical memory hierarchy example is illustrated in Fig. 1.1a. There are many policies to utilize this memory hierarchy efficiently, but generally speaking, we use a faster memory to store frequently accessed data to take advantages of its high speed, and overcome its capacity limitation by moving less frequently accessed data down to a slower memory. As a result, this memory hierarchy enables the CPU to exploit the low latency of the L1 SRAM cache as well as the large capacity of the HDD at the same time.

HEES systems aim at similar benefits by using multiple heterogeneous EES elements. Instead of relying on a single type of EES element, HEES systems exploit distinct advantages of multiple heterogeneous EES elements and hide their drawbacks. For instance, electric vehicles (EVs) and hybrid EVs (HEVs) exhibit frequent charging and discharging cycles with a short period at a high current rate. Conventional batteries make it difficult to maintain a high efficiency and a longer cycle life in such an operational environment. Replacing the batteries with supercapacitors may be huge upgrade in terms of efficiency and cycle life. However, current supercapacitor technologies have serious disadvantages in energy density and cost, which makes it impractical to completely replace the large battery in (H)EVs with supercapacitors. Instead, the use of supercapacitors in a complementary manner reinforces the drawback of battery through high power density, long life cycles, and high efficiency [2,4,5,10].

A conceptual drawing of a HEES system is shown in Fig. 1.1b. A HEES system is comprised of a number of EES banks, and is connected to external power sources and load devices. The HEES system in Fig. 1.1b is composed of a supercapacitor, Li-ion battery, and lead-acid battery. Similar to the computer memory hierarchy, the HEES system exploits different superiorities of these three energy storages: high power density and long life of the supercapacitor and the relatively low cost and high energy density of Li-ion battery and lead-acid battery. *Charge transfer interconnect (CTI)* internally connects the energy storages, external power sources, and external load devices though appropriate power converters.

Employing a HEES system concept comes with additional design considerations. Deployment of an HEES system does not always guarantee better performance unless proper design consideration is elaborated. Designers should carefully determine the selection of EES elements, capacity proportion of each EES element, system architecture, management policy, etc., in order to maximize the benefits of HEES system over the conventional homogeneous EES systems. There are different levels of approaches, from material development for EES elements to high-level system management, to maximize energy efficiency. This book is to leverage the advantages of HEES systems with elaborated architecture design and control policies mainly focusing on improving the system-level energy efficiency. More specifically, we consider the following questions throughout this book.

- How to leverage the advantages of different EES elements?
- What factors and components affect the energy efficiency of the HEES systems?
- How to maximize the energy efficiency of HEES systems through architecture-level and system-level design methodologies?
- How to achieve the joint optimization of HEES systems together with the power sources and load devices?

Contributions of this book can be summarized as follows.

- We propose new architecture designs for the HEES systems to maximize the energy efficiency. We introduce optimization issues involved in the new architecture designs and their operations.
- We model optimization problems of HEES systems as existing EDA problems and utilize the solution methods for the EDA problems for the HEES systems. We show the potential that systematic optimization methods of EDA problems can be used for the HEES system optimizations.
- We study limitations of conventional operating methods and suggests a new method that maximizes energy efficiency. We specially focus on maximizing solar energy harvesting considering the energy efficiency of the HEES systems.
- We implement a prototype of a HEES system to verify the proposed control methods.

Chapter 2 reviews the background study on the EES elements and related work in EES and HEES systems. Chapter 3 is a discussion on the architecture, components, and design considerations of the HEES systems. Chapter 4 introduces two novel architecture designs to improve energy efficiency. Chapter 5 expands the optimization scope to include the power sources. Chapter 6 introduces some experimental results and HEES system prototype implementation. Chapter 7 concludes this book and suggests future research directions.

References

1. Cleveland CJ (1992) Energy quality and energy surplus in the extraction of fossil fuels in the U.S. Ecological Economics 6(2):139–162
2. Frenzel B, Kurzweil P, Rönnebeck H (2011) Electromobility concept for racing cars based on lithium-ion batteries and supercapacitors. Journal of Power Sources 196(12):5364–5376
3. Koushanfar F (2010) Hierarchical hybrid power supply networks. In: Proceedings of the ACM/IEEE Design Automation Conference (DAC), pp 629–630
4. Lukic S, Wirasingha S, Rodriguez F, Cao J, Emadi A (2006) Power management of an ultracapacitor/battery hybrid energy storage system in an HEV. In: Proceedings of the IEEE Vehicle Power and Propulsion Conference (VPPC), pp 1–6
5. Moreno J, Ortuzar M, Dixon J (2006) Energy-management system for a hybrid electric vehicle, using ultracapacitors and neural networks. IEEE Transactions on Industrial Electronics 53(2):614–623
6. Odum HT (1975) Energy quality and carrying capacity of the earth. Tropical Ecology 16(1):14
7. Odum HT (1988) Self-organization, transformity, and information. Science 242(4882): 1132–1138
8. Park C, Seo J, Seo D, Kim S, Kim B (2003) Cost-efficient memory architecture design of nand flash memory embedded systems. In: Proceedings of the International Conference on Computer Design (ICCD), pp 474–480
9. Pedram M, Chang N, Kim Y, Wang Y (2010) Hybrid electrical energy storage systems. In: Proceedings of the ACM/IEEE International Symposium on Low-Power Electronics and Design (ISLPED), pp 363–368
10. Thounthong P, Raël S, Davat B (2009) Energy management of fuel cell/battery/supercapacitor hybrid power source for vehicle applications. Journal of Power Sources 193(1):376–385

Chapter 2
Background and Related Work

2.1 Electrical Energy Storage Elements

An *EES element* is a unit device or apparatus that stores electrical energy. Electrical energy can be stored in various forms of energy, such as mechanical energy, thermal energy, electrochemical energy, electrostatic energy, etc. Each of them has distinctive characteristics and applications. Among them, we focus on batteries and supercapacitors in this book. Those are the most widely deployed types of EES elements for various applications from small portable devices to grid-scale EES systems.

Common desirable characteristics of EES elements are as follows:

- **High cycle efficiency:** Cycle efficiency of an EES element is defined as the ratio of the amount of energy output during discharging to the amount of energy input during charging. A high cycle efficiency close to 100 % implies that less energy is lost during charging and discharging cycling.
- **Long cycle life:** Cycle life is the maximum number of charging and discharging cycles that an EES element can perform before its capacity drops below a specific percentage. After the cycle life is over, EES elements need to be replaced.
- **Low self-discharge rate:** Self-discharge or leakage is a measure of the speed that an EES element lose its energy even there is no current consumed by a load. For a long-term energy storage, low self-discharge EES elements are preferred.
- **High energy and power density:** Energy density is maximum energy storage per volume or weight. Power density is maximum power rating per volume or weight. For automotive or portable applications where volume and weight are critical constraints, high energy and power densities are important.
- **Low capital cost:** In order to meet energy and power requirements with limited capital constraint, an EES element with low cost per unit energy and per unit power is preferred.

Y. Kim, N. Chang, *Design and Management of Energy-Efficient Hybrid Electrical Energy Storage Systems*, DOI 10.1007/978-3-319-07281-4_2

Table 2.1 Comparison of EES elements [10, 12, 21, 24, 28, 46, 56]

EES elements	Power density (W/kg)	Power density (Wh/kg)	Capital cost ($/kWh)	Cycle efficiency (%)	Cycle life	Self-discharge per day
Lead-acid battery	180	30–40	74–222	70–90	500–800	0.1–0.3 %
Li-ion battery	1,800	150–250	1,040–1,484	80–90	1,200	0.1–0.3 %
NiMH battery	250–1,000	30–80	450–1,000	66	500–1,000	2 %
NiCd battery	150	40–60	296–890	70–90	1,500	0.2–0.6 %
Metal-air battery	–	450–650	74–296	<50	100+	Very small
Super capacitor	1,000–2,000	2.5–15	2,000	>93	100,000+	20–40 %

Table 2.1 compares some important characteristics of several representative batteries and supercapacitor. They all have distinctive advantages and weaknesses. For example, metal-air batteries have an outstanding energy density, but cycle efficiency is poor. On the other hand, supercapacitors have a superior power density and cycle efficiency, but their energy density is low and they suffer from high self-discharge. None of them has superior characteristic in all of the metrics, as we discussed in Chap. 1 as the motivation of HESS systems.

Depending on their characteristics, each EES element has typical applications. For instance, lead-acid batteries have a relatively low cost per energy, and so they are suitable for large-scale energy storages. Li-ion batteries are widely used for mobile and automotive applications thanks to their high energy density and long cycle life. Supercapacitors are utilized in applications that require high power capability for a short duration. More discussion and comparison of batteries and supercapacitor can be found from [10, 12, 21, 24, 28, 46, 56].

2.2 Previous Electrical Energy Storage Systems

2.2.1 System Architectures

A typical EES system is a homogeneous EES system that consists of a single type of EES elements. This is natural since the homogeneity offers ease of implementation, control and maintenance. Figure 2.1 illustrates the architecture of a homogeneous EES system. It is composed of EES elements, input power converter, and output power converter. Typically, in order to increase energy capacity and/or power capacity, multiple EES elements are clustered. The input converter performs energy transduction and/or power regulation from the power source (e.g., power grid, solar

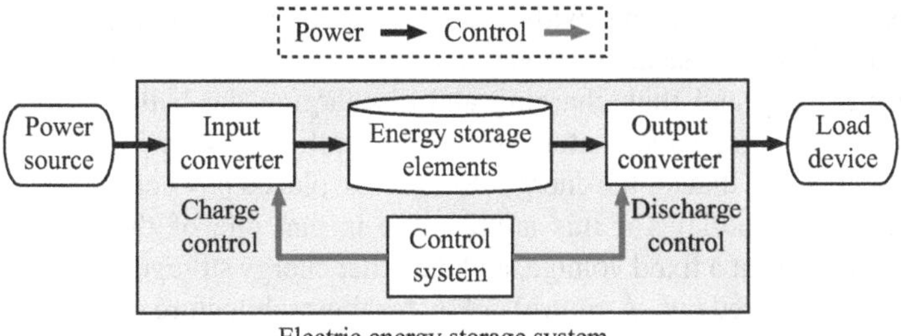

Fig. 2.1 Architecture of a typical homogeneous EES system

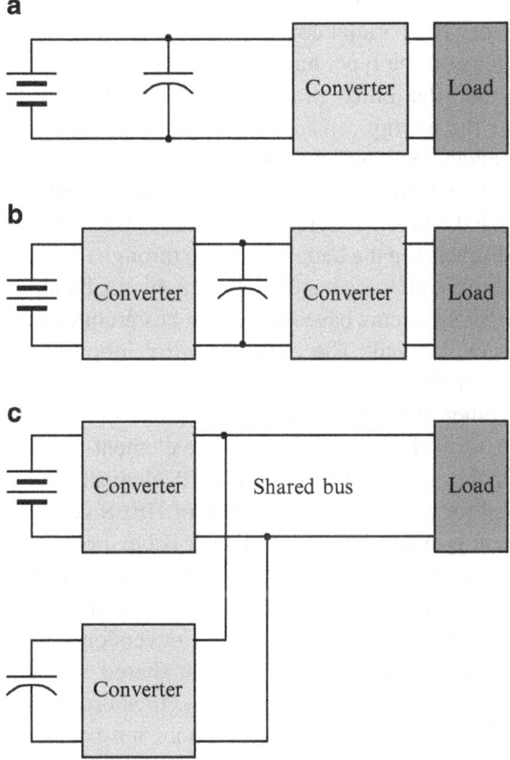

Fig. 2.2 Battery-supercapacitor hybridization topologies. (**a**) Passive parallel connection topology. (**b**) Cascade converters topology. (**c**) Shared bus topology

photovoltaic panels, generators, etc.) to EES elements, and the output converter performs the same from the EES elements to load devices (e.g., electric appliances, EV motors, etc.).

The system architecture of a HEES system is more complicated. Figure 2.2 shows some HEES architectures proposed in previous literature. Some HEES architectures are specialized for particular purposes, such as lowering the effective internal resistance by parallel connection of battery and supercapacitor (Fig. 2.2a) [17], or buffering battery current with supercapacitors through a power converter (Fig. 2.2b) [44, 45]. Cascaded converter architecture in Fig. 2.2b has an explicit physical hierarchy and thus charging and discharging the battery should always be buffered by the supercapacitor. Shared bus architecture in Fig. 2.2c is a more general architecture that all the energy storages are placed physically flat [49,50,55]. A typical control policy for this architecture is that one of them maintains the shared bus voltage at a fixed voltage, and the other energy storages inject designated current into the shared bus. A control policy for this architecture is to regulate a fast-responding EES element's voltage with a slow-responding EES element [49, 50]. Another control method is to increase the supercapacitor current as the battery current increases [55].

However, the previous HEES systems cannot always fully exploit the advantages of the HEES systems. The parallel connection and cascaded converter architectures have strict limitations on the type, number, and topology of EES elements, and so lack of scalability and flexibility. In addition, the parallel connection architecture cannot fully utilize the energy capacity of the supercapacitor because its state-of-charge (SoC)-dependent voltage cannot deviate from the small battery terminal voltage variation. It also does not allow independent current control of each EES element. The cascaded converter architecture is subject to a lower efficiency because the charging and discharging the battery must go through two conversion steps even if the current is low enough to be handled directly by the battery.

The previous HEES systems based on shared bus architecture are not optimized for energy efficiency as well. The energy control loops do not allow arbitrary current flow for each EES bank. Current distribution among the EES elements is determined by other storage element's voltage or current. Therefore, it is not possible to charge or discharge the each storage elements with the energy-optimal current considering the characteristics of each EES element. Also, the energy control loop is for designed for a specific combination of HEES systems, and the behavior of each EES element is coupled with the state or behavior other EES elements. The whole energy control loop is based on pre-characterization, and not scalable nor flexible. Furthermore, it is often neglected that the power converter is a significant contributor to system-level energy efficiency. Conversion efficiency in the shared bus architecture is not always maximum if the shared bus voltage is fixed, and this may result in sub-optimal energy efficiency. In short, the energy control loop of previous shared bus-based HEES systems does not consider important factors related with the system-level energy efficiency

2.2.2 *Applications of EES Systems*

EES systems have been widely utilized for various applications. A comprehensive review on various EES system applications is presented in [10], but the focus of this review is mainly on large grid-scale applications. We expand the scope to medium- and small-scale applications, and we also discuss how HEES approach can be utilized in those application areas.

2.2.2.1 Power Grid Applications

Many EES systems have been practically deployed for various applications in the power grid [10, 15]. They can be used as a commodity storage for storing energy during off-peak period for the use during peak period for arbitraging electricity price. Such applications may decrease peak load demand (i.e., peak shaving), or make the load demand uniform over time (i.e., load leveling). The purpose of such applications is mainly reducing the maximum capacity requirement in generation, transmission, and distribution systems, and hence reduce the cost of power generation.

Another application of EES systems is contingency service, which supplies power when grid power generation plants fall off-line, to provide power without interruption. Figure 2.3 is a grid-connected NiCd battery-based EES system operated by Golden Valley Electric Association (GVEA), Alaska, which is designed to provide 15 min of community load against the power failure [42].

Another important role of EES systems is frequency regulation. Frequency regulation is short-term balancing of power generation and demand in order to maintain a fixed AC frequency. EES systems are capable of modulating their power generation or consumption at a high speed, and so they are suitable for performing frequency regulation [20, 29, 39, 53]. Recently, grid-connected HEES systems are investigated for frequency regulation applications. UltraBattery technology is a battery-supercapacitor HESS of passive parallel hybridization architecture. A grid-connected HEES system composed of UltraBattery is built and tested for frequency regulation [54]. A genetic algorithm-based revenue maximization method for HEES systems used in energy and regulation markets is proposed in [26].

Fig. 2.3 A 27 MW NiCd battery EES system for power grid by Golden Valley Electric Association EES system, Fairbanks, AK

EES systems can be installed on the consumer side for stable and economic electricity supply. There are some high-level control methods developed for Li-ion battery-based EES systems for residential purpose to reduce system capital cost and energy cost by peak shaving and load leveling when time-of-use charge is applied [13, 41]. Another management method introduced employs dynamic programming and expert knowledge base rules to reduce capital cost and energy cost for industrial purpose [31].

2.2.2.2 Renewable Power Generation

For reliable and efficient power generation from renewable power sources, such as solar cells and windmills, EES systems are often mandatory. The level of power generation of such power sources largely determined by uncontrollable environmental factors. The use of EES systems have two major benefits for renewable power generation. First, an EES system can increase energy utilization by mitigating the temporal mismatch between the power generation and load demand. If load demand is lower then maximum power generation capability, excess power that is not stored is wasted and cannot be used later. By storing the excess power and using it when load demand is higher than maximum power generation capability, we can reduce energy waste and fully utilize the power generation capability always [3, 5, 12, 18, 19, 41, 48]. Second, an EES system improves power generation stability. Due to highly variable power generation depending on the environmental changes, instantaneous power supply-demand mismatch results in severe variations in frequency [6, 11]. EES systems can maintain a desired frequency by rapidly modulating its power supply or draw in response to the frequency variations.

2.2.2.3 Electric Vehicles

EVs are an active application area of battery-based EES systems today. Hybrid EVs (HEVs), which are powered by both batteries and an internal combustion engine (ICE), are gaining popularity due to their high fuel efficiency. In a HEV, it is important to determine the power distribution between batteries and ICE. Various optimization techniques, such as reinforcement learning and dynamic programming, can be used, and sometimes prediction of future driving cycles is utilized for the optimization [8, 23, 32, 37]. Recently, many researches propose to use a HEES system, instead of a simple battery-based EES system, to further improve energy efficiency [2,22,27,36,38,43,49]. When a HEES system in used, however, the power distribution problem becomes more complex. A power control method introduced in [49] is for HEVs with a shared bus hybridization architecture, shown in Fig. 2.2c, composed of a fuel cell, battery, and supercapacitor. A power control method proposed in [36] optimizes the supercapacitor current using neural network and achieves more than 20 % of km/kWh improvement.

Fig. 2.4 An example of battery-based EES system applications, Toyota plug-in hybrid vehicle with Li-ion batteries

Supercapacitors also can be utilized as a high-power and high-efficiency energy buffer for acceleration and regenerative braking [16]. In [9], a supercapacitor-only EV, which has no other power sources like an ICE or battery, is introduced. However, due to the limited energy density of supercapacitor, it is practical to use them together with batteries or ICE (Fig. 2.4).

Another research direction for EES systems in EVs is improving their long-term economic feasibility, such as cost reduction and cycle life maximization [4]. Economic viability of using the HEES system for EVs is analyzed in [34].

2.2.2.4 High-Power Machineries and Tools

Supercapacitors are attracting more attentions for their high power density, long life cycles, and high efficiency. An article introduces and categorizes various applications of supercapacitor-based EES systems [35]. Some applications can take advantages the high power capacity and high cycle life of supercapacitors, for example, heavy industrial machineries and heavy vehicles as shown in Fig. 2.5a. Power flow in those applications shows frequent charge/discharge cycles with a short period and a large amount of current, which make supercapacitors beneficial. Day-night storages, which are charged during daytime and discharged during nighttime, are an application that utilizes its high cycle efficiency and long cycle life. Their fast-charging advantage can be utilized for home-use power tools such as a cordless screwdriver as shown in Fig. 2.5b.

2.2.2.5 Low-Power Embedded Systems

For low-power applications, high cycle efficiency of supercapacitors is beneficial. Energy harvesting is common for self-sustainability for sensor nodes, and

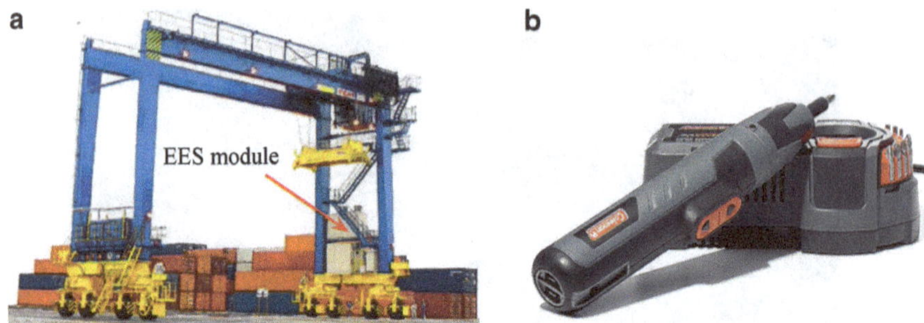

Fig. 2.5 Examples of supercapacitor-based EES system applications [35]. (a) Regenerative energy storage for a seaport crane. (b) Coleman FlashCell cordless screwdriver

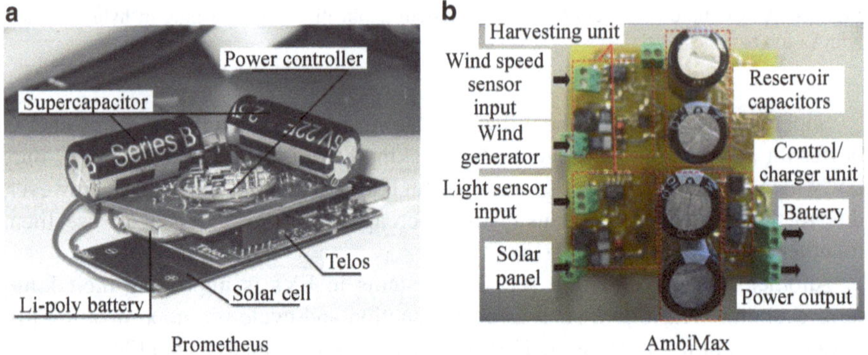

Fig. 2.6 Battery-supercapacitor hybrid wireless sensor nodes. (a) Prometheus from UC Berkeley [25] and (b) AmbiMax from UC Irvine [40]

supercapacitors provide a high cycle efficiency for the harvested energy. Examples in literatures are solar energy harvesting [47] and vibration energy harvesting [51]. Some low-power sensor nodes like [25, 40] employ a battery-supercapacitor hybrid as shown Fig. 2.6. Due to very limited capability to produce power from energy harvesting devices such as solar cells, reducing power loss during charge/discharge cycles is important. They take advantages of the high cycle efficiency of supercapacitors while using a battery as a low-leakage long-term energy storage.

2.2.2.6 Other EES Systems

Other than the battery-based and supercapacitor based EES systems we discussed in this section, there are more EES systems based on different energy storage mechanisms, such as kinetic, thermal and chemical energy storages are also available. Examples include flywheels [1], compressed air [33], hydropower using dams [14], thermal energy storage system [30], and hydrogen-based chemical energy storage system [7]. Although we do not cover them in this book, one can refer to [10, 52] for other types of energy storage systems.

References

1. Akagi H, Sato H (2002) Control and performance of a doubly-fed induction machine intended for a flywheel energy storage system. IEEE Transactions on Power Electronics 17(1):109–116
2. Allegre A, Bouscayrol A, Trigui R (2009) Influence of control strategies on battery/supercapacitor hybrid energy storage systems for traction applications. In: Proceedings of the IEEE Vehicle Power and Propulsion Conference (VPPC), pp 213–220
3. Barote L, Weissbach R, Teodorescu R, Marinescu C, Cirstea M (2008) Stand-alone wind system with vanadium redox battery energy storage. In: Proceedings of the International Conference on Optimization of Electrical and Electronic Equipment (OPTIM), pp 407–412
4. Bashash S, Moura SJ, Forman JC, Fathy HK (2011) Plug-in hybrid electric vehicle charge pattern optimization for energy cost and battery longevity. Journal of Power Sources 196(1):541–549
5. Bernal-Agustn, Dufo-Lpez R (2009) Simulation and optimization of stand-alone hybrid renewable energy systems. Renewable and Sustainable Energy Reviews 13(8):2111–2118
6. Bevrani H, Ghosh A, Ledwich G (2010) Renewable energy sources and frequency regulation: survey and new perspectives. IET Renewable Power Generation 4(5):438–457, DOI 10.1049/iet-rpg.2009.0049
7. Bilodeau A, Agbossou K (2006) Control analysis of renewable energy system with hydrogen storage for residential applications. Journal of Power Sources
8. Borhan H, Vahidi A, Phillips A, Kuang M, Kolmanovsky I, Di Cairano S (2012) MPC-based energy management of a power-split hybrid electric vehicle. IEEE Transactions on Control Systems Technology 20(3):593–603, DOI 10.1109/TCST.2011.2134852
9. Carter P, Baxter J, Newill T, Erekson T (2005) An ultracapacitor-powered race car update. In: Proceeding of the Electrical Insulation Conference and Electrical Manufacturing Expo, pp 267–274
10. Chen H, Cong TN, Yang W, Tan C, Li Y, Ding Y (2009) Progress in electrical energy storage system: A critical review. Progress in Natural Science 19(3):291–312
11. Cho Y, Shim JW, Kim SJ, Min SW, Hur K (2013) Enhanced frequency regulation service using hybrid energy storage system against increasing power-load variability. In: IEEE PES
12. Chowdhury M, Haque M, Aktarujjaman M, Negnevitsky M, Gargoom A (2011) Grid integration impacts and energy storage systems for wind energy applications—a review. In: Proceedings of the IEEE Power and Energy Society General Meeting, pp 1–8
13. Darcovich K, Gupta N, Davidson I, Caroni T (2010) Residential electrical power storage scenario simulations with a large-scale lithium ion battery. Journal of Applied Electrochemistry 40:749–755
14. Deane JP, Gallachóir BPÓ, McKeogh E (2010) Techno-economic review of existing and new pumped hydro energy storage plant. Renewable and Sustainable Energy Reviews 14(4):1293–1302
15. Divya K, Østergaard J (2009) Battery energy storage technology for power systems—an overview. Electric Power Systems Research 79(4):511–520
16. Dixon J, Ortuzar M (2002) Ultracapacitors + dc-dc converters in regenerative braking system. IEEE Aerospace and Electronic Systems Magazine 17(8):16–21, DOI 10.1109/MAES.2002.1028079
17. Dougal R, Liu S, White R (2002) Power and life extension of battery-ultracapacitor hybrids. IEEE Transactions on Components and Packaging Technologies 25(1):120–131

18. Ekren O, Ekren BY (2010) Size optimization of a PV/wind hybrid energy conversion system with battery storage using simulated annealing. Applied Energy 87(2):592–598
19. Ekren O, Ekren BY, Ozerdem B (2009) Break-even analysis and size optimization of a PV/wind hybrid energy conversion system with battery storage – a case study. Applied Energy 86(7–8):1043–1054
20. of Energy UD (2013) Grid energy storage
21. Evans A, Strezov V, Evans TJ (2012) Assessment of utility energy storage options for increased renewable energy penetration. Renewable and Sustainable Energy Reviews 16(6):4141–4147
22. Frenzel B, Kurzweil P, Rönnebeck H (2011) Electromobility concept for racing cars based on lithium-ion batteries and supercapacitors. Journal of Power Sources 196(12):5364–5376
23. He Y, Chowdhury M, Pisu P, Ma Y (2012) An energy optimization strategy for power-split drivetrain plug-in hybrid electric vehicles. Transportation Research Part C: Emerging Technologies 22(0):29–41, DOI http://dx.doi.org/10.1016/j.trc.2011.11.008, URL http://www.sciencedirect.com/science/article/pii/S0968090X11001604
24. Hou Y, Vidu R, Stroeve P (2011) Solar energy storage methods. Industrial & Engineering Chemistry Research 50(15):8954–8964
25. Jiang X, Polastre J, Culler D (2005) Perpetual environmentally powered sensor networks. In: Proceedings of the International Symposium on Information Processing in Sensor Networks (IPSN), pp 463–468
26. Jin C, Lu S, Lu N, Dougal R (2011) Cross-market optimization for hybrid energy storage systems. In: Proceedings of the IEEE Power and Energy Society General Meeting, pp 1–6
27. Khaligh A, Li Z (2010) Battery, ultracapacitor, fuel cell, and hybrid energy storage systems for electric, hybrid electric, fuel cell, and plug-in hybrid electric vehicles: State of the art. IEEE Transactions on Vehicular Technology 59(6):2806–2814
28. Kim Y, Wang Y, Chang N, Pedram M (2013) Computer-aided design and optimization of hybrid energy storage systems. Foundations and Trends in Electronic Design Automation 7(5):247–338
29. Kirby BJ (2004) Frequency regulation basics and trends. Tech. rep., Oak Ridge National Laboratory
30. LeBreux M, Lacroix M, Lachiver G (2009) Control of a hybrid solar/electric thermal energy storage system. International Journal of Thermal Sciences pp 645–654
31. Lee TY, Chen N (1995) Determination of optimal contract capacities and optimal sizes of battery energy storage systems for time-of-use rates industrial customers. IEEE Transactions on Energy Conversion 10(3):562–568
32. Liu J, Peng H (2008) Modeling and control of a power-split hybrid vehicle. IEEE Transactions on Control Systems Technology 16(6):1242–1251, DOI 10.1109/TCST.2008.919447
33. Lund H, Salgi G (2009) The role of compressed air energy storage (CAES) in future sustainable energy systems. Energy Conversion and Management 50(5):1172–1179
34. Miller J, Deshpande U, Dougherty T, Bohn T (2009) Power electronic enabled active hybrid energy storage system and its economic viability. In: Proceedings of the IEEE Applied Power Electronics Conference and Exposition (APEC), pp 190–198
35. Miller JR, Burke AF (2008) Electrochemical capacitors: Challenges and opportunities for real-world applications. The Electrochemical Society Interface 17:53–57
36. Moreno J, Ortuzar M, Dixon J (2006) Energy-management system for a hybrid electric vehicle, using ultracapacitors and neural networks. IEEE Transactions on Industrial Electronics 53(2):614–623
37. Moura S, Fathy H, Callaway D, Stein J (2011) A stochastic optimal control approach for power management in plug-in hybrid electric vehicles. IEEE Transactions on Control Systems Technology 19(3):545–555, DOI 10.1109/TCST.2010.2043736
38. Ortuzar M, Moreno J, Dixon J (2007) Ultracapacitor-based auxiliary energy system for an electric vehicle: Implementation and evaluation. IEEE Transactions on Industrial Electronics 54(4):2147–2156
39. Oudalov A, Chartouni D, Ohler C (2007) Optimizing a battery energy storage system for primary frequency control. IEEE Trans Power Syst

40. Park C, Chou P (2006) AmbiMax: Autonomous energy harvesting platform for multi-supply wireless sensor nodes. In: Proceedings of the IEEE Communications Society Conference on Sensor, Mesh and Ad Hoc Communications and Networks (SECON), vol 1, pp 168–177
41. Park S, Wang Y, Kim Y, Chang N, Pedram M (2012) Battery management for grid-connected pv systems with a battery. In: Proceedings of the ACM/IEEE International Symposium on Low-Power Electronics and Design (ISLPED), pp 115–120
42. Roberts B, McDowall J (2005) Commercial successes in power storage. IEEE Power and Energy Magazine 3(2):24–30
43. Shah V, Chaudhari R, Kundu P, Maheshwari R (2010) Performance analysis of hybrid energy storage system using hybrid control algorithm with bldc motor driving a vehicle. In: Proceedings of the Joint International Conference on Power Electronics, Drives and Energy Systems (PEDES), pp 1–5
44. Shin D, Kim Y, Seo J, Chang N, Wang Y, Pedram M (2011) Battery-supercapacitor hybrid system for high-rate pulsed load applications. In: Proceedings of the Design, Automation & Test in Europe Conference & Exhibition (DATE), pp 1–4
45. Shin D, Kim Y, Wang Y, Chang N, Pedram M (2012) Constant-current regulator-based battery-supercapacitor hybrid architecture for high-rate pulsed load applications. Journal of Power Sources 205(0):516–524
46. Shkolnikov E, Zhuk A, Vlaskin M (2011) Aluminum as energy carrier: Feasibility analysis and current technologies overview. Renewable and Sustainable Energy Reviews 15(9):4611–4623
47. Simjee F, Chou P (2006) Everlast: Long-life, supercapacitor-operated wireless sensor node. In: Proceedings of the ACM/IEEE International Symposium on Low-Power Electronics and Design (ISLPED), pp 197–202
48. Teleke S, Baran M, Bhattacharya S, Huang A (2010) Optimal control of battery energy storage for wind farm dispatching. IEEE Transactions on Energy Conversion 25(3):787–794
49. Thounthong P, Raël S, Davat B (2009) Energy management of fuel cell/battery/supercapacitor hybrid power source for vehicle applications. Journal of Power Sources 193(1):376–385
50. Thounthong P, Chunkag V, Sethakul P, Sikkabut S, Pierfederici S, Davat B (2011) Energy management of fuel cell/solar cell/supercapacitor hybrid power source. Journal of Power Sources 196(1):313–324
51. Torah R, Glynne-Jones P, Tudor M, O'Donnell T, Roy S, Beeby S (2008) Self-powered autonomous wireless sensor node using vibration energy harvesting. Measurement Science and Technology 19(12):125,202–125,209
52. Vazquez S, Lukic S, Galvan E, Franquelo L, Carrasco J (2010) Energy storage systems for transport and grid applications. IEEE Transactions on Industrial Electronics 57(12):3881–3895, DOI 10.1109/TIE.2010.2076414
53. Walawalkar R, Apt J, Mancini R (2007) Economics of electric energy storage for energy arbitrage and regulation in New York. Energy Policy
54. Wood J (2012) Integrating renewables into the grid: Applying UltraBattery® technology in MW scale energy storage solutions for continuous variability management. In: IEEE POWERCON
55. Zhang Y, Jiang Z, Yu X (2008) Control strategies for battery/supercapacitor hybrid energy storage systems. In: Proceedings of the IEEE Energy 2030 Conference, pp 1–6
56. Zhao X, Sanchez BM, Dobson PJ, Grant PS (2011) The role of nanomaterials in redox-based supercapacitors for next generation energy storage devices. Nanoscale 3:839–855

Chapter 3
Hybrid Electrical Energy Storage Systems Design

In this chapter, we discuss high-level concepts of HEES systems. We first present the desirable characteristics of a HEES system that we achieve by the optimization techniques that we discuss in this book. We next present the architecture of a HEES system that we consider throughout this book, and explain its components.

3.1 Design Considerations of HESS Systems

Designing a HEES system requires optimization effort in different levels from the material level (e.g., battery electrolyte materials) to the system level (e.g., state-of-health (SoH) management). In this book, we focus on system-level design and optimization techniques. From the system-level point of view, we emphasize the following aspects 31233 of HEES systems.

- Scalability: A HEES system architecture should be able to accommodate increased number of EES banks.
- Modularity: Adding, removing, or modifying configuration of EES elements should be easy, without a significant modification to the whole HEES system.
- Flexibility: A HEES system can adopt various types of EES elements, power sources, and load devices.

These metrics are often neglected in conventional EES system design because the homogeneity makes the architecture design and control straightforward. However, they should not be overlooked when designing and implementing HEES systems, which adopt a sophisticated architecture and involve complicated system-level controls for heterogeneous EES elements.

Y. Kim, N. Chang, *Design and Management of Energy-Efficient Hybrid Electrical Energy Storage Systems*, DOI 10.1007/978-3-319-07281-4_3

3.2 HESS System Architecture

In order to ensure scalability, modularity, and flexibility, we propose a system architecture as shown in Fig. 3.1. The HEES system is composed of multiple, heterogeneous *EES banks*. Each EES bank consists of an *EES array* and a bidirectional power converter that charges and discharges the EES array. The HEES system also has unidirectional power converters to connect the HEES system to various kinds of AC or DC power sources and load devices. The CTI is an interconnection network for charge transfers among EES banks, power sources, and load devices through power converters. We discuss these system components in Sect. 3.4 in more detail.

We employ physically a flat architecture for flexibility; there is no physical hierarchy among EES banks. The CTI can be a single shared bus similar to Fig. 2.2c, or other topologies, such as multiple buses, segmented buses, mesh interconnect, crossbar interconnect, or any combinations of those. Of course, although there is no physical hierarchy, system-level management policies should exploit logical hierarchy among EES banks.

The system controller performs high-level system control and management for reliable and energy-efficient operations. It is in charge of determining the CTI voltage and current of each EES banks and power sources based on the load current and EES bank status such as SoC and SoH. While the system controller makes high-level decisions with a software control loop, power converters maintain CTI voltage and input/output current determined by the controller with a hardware feedback control loop. The outer control loop continuously updates the set points to maximize the system efficiency by the use of high-level management policies. We are able to maintain a reasonable set point update frequency free from CTI stability

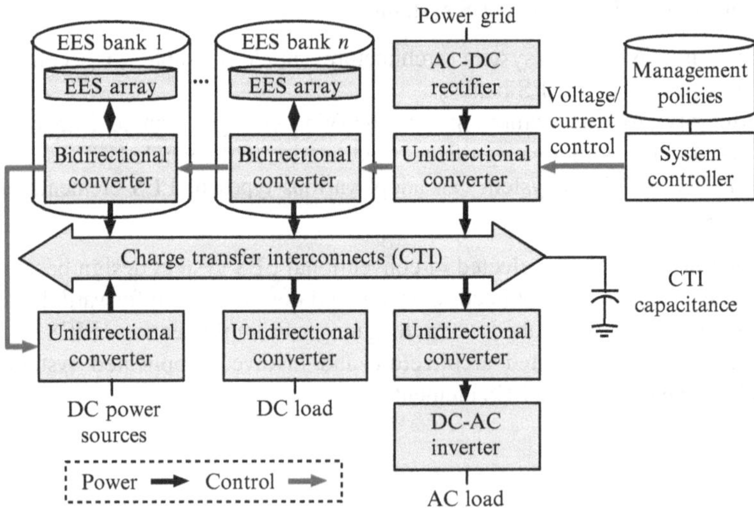

Fig. 3.1 Architecture of the proposed HEES system

control thanks to the cascaded feedback loops, which allows us to use standard microcontroller for the outer feedback loop. Detailed design and implementation will be explained in Sect. 6.2.

3.3 Charge Transfer and Charge Management

The benefits of the HEES systems over the conventional EES systems rely on sophisticated charge management. *Charge management* is optimizing the charge transfer among the EES nodes. It includes determining the optimal time and amount of charge transfers to each EES nodes and finding the optimal operating conditions of the charge transfers. Compared with the homogeneous EES systems where the charge and discharge can be distributed uniformly across the all the homogeneous cells, charge management in the HEES systems is not straightforward.

We should select particular EES banks to charge or discharge among multiple possible selections and determine the CTI voltage and amount of current that maximize the energy efficiency. Also, we may need to internally move energy from one EES bank to another in order to mitigate self-discharge or to prepare for expected demand for energy/power capacity. For the last few years, we have proposed various charge management policies for HEES systems. Our charge management policies include (i) charge allocation for charging EES banks [9, 13], (ii) charge replacement for discharging EES banks [11, 12], and (iii) charge migration for moving energy between EES banks [7, 8]. We consider the characteristics of the EES elements, power converter efficiency, input or output power variations, and time constraint, and find the EES banks and amount of current that achieves the energy-optimal charge transfers.

Once the sources and destinations of the charge transfers are determined by the charge management policies, *charge transfer scheduling* follows [14]. Charge transfer scheduling is to determine the duration of the CTI occupation of each charge transfer. It derives the charge current that determines the time duration required to finish the charge transfer. They should be determined carefully not only to maximize the efficiency of a single charge transfer, but also considering efficiency of other charge transfers.

3.4 HESS System Components

3.4.1 Nodes

A *node* in a HEES system is entity that generates, stores, or consumes energy. There are multiple nodes in a HEES system, and they include EES banks, power sources, and load devices.

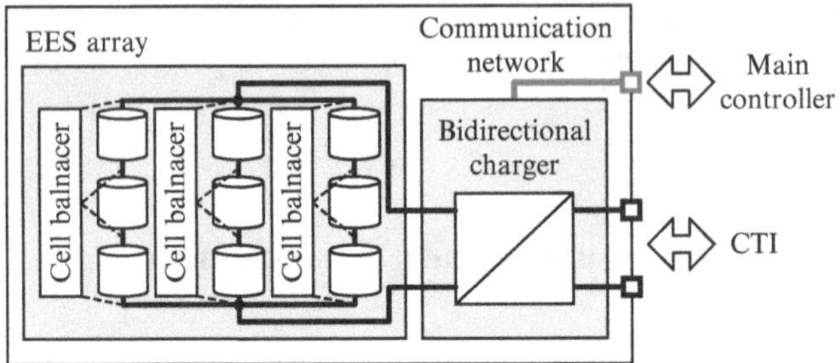

Fig. 3.2 EES bank with a 3 × 3 EES array and charger

3.4.1.1 Electrical Energy Storage Banks

Each bank consists of homogeneous EES elements. All the EES bank should meet standard interface specifications though the inside EES element array is heterogeneous. The standard specification is defined at the bank terminal such as terminal voltage range and the communication network protocol. The maximum charging and discharging current is generally different by EES bank to exploit the hybrid concept. The mandatory component in an EES bank includes a homogeneous EES element array, a bidirectional charger and a communication network.

The EES array is a set of multiple identical EES elements that are connected in series and/or parallel forming an $n \times m$ regular matrix, where n is the number of series connections and m is the number of parallel connections. The dimension of the EES element array is determined by the power and energy capacity, and the maximum voltage rating of the EES bank. We consider the regular array structure only so that we maintain the same SoC and SoH of all the elements in the array. We may also consider reconfigurable architecture of the EES array [2], but we do not consider the reconfiguration for the prototype implementation because the dimension of the EES array is not large enough for reconfiguration. Figure 3.2 shows an HEES bank architecture with a 3 × 3 array for an illustration purpose.

The charger regulates the current to and from the EES bank array and CTI. The bidirectional charger can also be set to a DC–DC converter when the direction is from the EES bank array to CTI. This allows easy CTI voltage control. The bidirectional charger is connected to the main controller through a communication network. When the power is on, the main controller identifies the bank characteristics such as the type of EES bank, the maximum charging/discharging current, the current SoC, the current SoH, the EES element array terminal voltage, temperature of the cells, etc. The main controller continues monitoring the current SoC, the EES element array terminal voltage, temperature of the cells, etc.

We optionally use cell balancers when the EES elements require external cell balancing. Even though all the EES elements are of the same type, manufacturing variation in practice may result in imbalance of characteristics such as capacity and internal resistance that causes imbalanced SoC during operation and even damage to the elements [1]. Supercapacitors and Li-ion batteries require external cell balancing while lead-acid batteries may work without external cell balancing for example.

3.4.1.2 Power Sources and Load Devices

The power converters should be designed considering the power input and output requirements. The grid-connected HEES system receives AC power from the power grid and supply AC power to the load devices. Batteries and supercapacitors are DC energy storages. Therefore, AC–DC rectifier is required to perform AC-to-DC conversion for charging the EES bank from the power grid, and DC–AC inverter is required to perform DC-to-AC conversion for discharging the EES bank to the load devices. These power converters for the power sources and load devices do not need to be bidirectional. They are connected to the CTI whose voltage is dynamically varying, and so they should be able to adapt to the voltage variation.

DC power sources such as fuel cells and solar cells require DC–DC conversion from the power sources to the CTI, and DC load devices such as portable appliances also require DC–DC conversion from the CTI to the load devices. Similar to the AC power converters, the DC–DC converters for the DC power sources and DC load devices are unidirectional, and they also need to be able to adapt to the CTI voltage variation.

3.4.2 Charge Transfer Interconnect

Similar to on-chip communication networks, the network topology of the CTI is one of the important design considerations for scalability and energy efficiency. The CTI architecture should be carefully determined for given type and number of EES banks. The parallel connection (Fig. 2.2a) and cascaded architecture (Fig. 2.2b) are not suitable for three or more EES banks. The shared bus architecture (Fig. 2.2c) provides higher scalability. The energy efficiency of the charge transfers is significantly affected by the CTI voltage level because the CTI is the input or output port of the power converters of each charger and the power converter efficiency is dependent on the input/output voltage and output current. Each charge transfer has its own optimal CTI voltage level that maximizes the energy efficiency [7–9, 11]. Therefore, the CTI voltage needs to be dynamically adjustable.

As the number of EES banks increases, more number of simultaneous charge transfers take places among the EES banks. Higher energy efficiency may be achieved if the CTI network is able to provide more isolated paths to the simultaneous charge transfers for the energy-optimal CTI voltage. The shared-bus architecture is the simplest design which is appropriate to accommodate a small number of EES nodes where there are not many simultaneous charge transfers. All the simultaneous charge transfers share the single CTI voltage level which may be energy-inefficient, but the single share bus offers easy implementation. We proposed more scalable architectures such as a mesh grid architecture [3] and multiple-bus architecture [14] for more number of EES banks. The mesh grid architecture uses routers like a communication network to dynamically reconfigure the CTI between the EES banks [3]. The routing algorithm merges charge transfers to increase routability and find the optimal CTI voltage to maximize the energy efficiency. EES banks in the multiple-bus architecture can be connected any of the multiple buses [14]. The charge transfers are merged into the same number of sets as the number of busses and each charge transfer set is assigned to one bus. Those architectures allow multiple simultaneous charge transfers at the energy-optimal CTI voltages levels, and therefore enhance the energy efficiency.

3.4.3 System Control and Communication Network

The HEES systems require sophisticated management policies than conventional EES systems do because of the heterogeneity of EES elements. Using multiple different EES elements doe not guarantee improved energy efficiency. Proper management policies are crucial for the HEES system to achieve energy efficiency improvement. It is mandatory to devices system-level policies in order for maximizing the benefits of the HEES system in energy efficiency, lifetime, etc., by exploiting its heterogeneity, which have been not considered for the conventional homogeneous EES systems.

In addition to the charge management policies introduced in Sect. 3.3, we proposed more HEES system optimization and management schemes that requires system-level controls. We also proposed bank reconfiguration to change the connection of EES elements within an EES bank to adjust to the optimal terminal voltage considering the CTI voltage and the power converter efficiency [2]. We also introduced a management scheme to enhance the SoH of batteries using supercapacitors for high frequency power detected by applying a crossover filter to the power profile [10].

These sophisticated management schemes require a high-level controller, rather than a simple feedback control loop. For example, the current from each EES bank is directly determined by the voltage or current of another EES bank [5, 6, 15] or speed of the EV [4]. However, taking non-linear and time-varying characteristics such as rate capability and cycle life into account requires a more elaborate control system. Also, the control system involves a considerable amount of control data transfer through a communication network for collecting information from many EES banks, power sources, and load devices and sending commands to them. High speed communication network is required to enable the high-speed control. In addition, scalability issue arises in the communication network like in the CTI architecture to accommodate increased number of EES banks. Therefore, the communication network should be designed considering the high-speed and scalability requirements.

References

1. Barsukov Y (2009) Battery cell balancing: what to balance and how. Tech. rep., Texas Instruments
2. Kim Y, Park S, Wang Y, Xie Q, Chang N, Poncino M, Pedram M (2011) Balanced reconfiguration of storage banks in a hybrid electrical energy storage system. In: Proceedings of the International Conference on Computer-Aided Design (ICCAD), pp 624–631
3. Kim Y, Park S, Chang N, Xie Q, Wang Y, Pedram M (2012) Networked architecture for hybrid electrical energy storage systems. In: Proceedings of the ACM/IEEE Design Automation Conference (DAC), pp 522–528
4. Moreno J, Ortuzar M, Dixon J (2006) Energy-management system for a hybrid electric vehicle, using ultracapacitors and neural networks. IEEE Transactions on Industrial Electronics 53(2):614–623
5. Thounthong P, Raël S, Davat B (2009) Energy management of fuel cell/battery/supercapacitor hybrid power source for vehicle applications. Journal of Power Sources 193(1):376–385
6. Thounthong P, Chunkag V, Sethakul P, Sikkabut S, Pierfederici S, Davat B (2011) Energy management of fuel cell/solar cell/supercapacitor hybrid power source. Journal of Power Sources 196(1):313–324
7. Wang Y, Kim Y, Xie Q, Chang N, Pedram M (2011) Charge migration efficiency optimization in hybrid electrical energy storage (hees) systems. In: Proceedings of the ACM/IEEE International Symposium on Low Power Electronics and Design (ISLPED), pp 103–108
8. Wang Y, Xie Q, Pedram M, Kim Y, Chang N, Poncino M (2012) Multiple-source and multiple-destination charge migration in hybrid electrical energy storage systems. In: Proceedings of the Design, Automation & Test in Europe Conference & Exhibition (DATE)
9. Xie Q, Wang Y, Kim Y, Chang N, Pedram M (2011) Charge allocation for hybrid electrical energy storage systems. In: Proceedings of the IEEE/ACM/IFIP International Conference on Hardware/Software Codesign and System Synthesis (CODES+ISSS), pp 277–284
10. Xie Q, Lin X, Wang Y, Pedram M, Shin D, Chang N (2012a) State of health aware charge management in hybrid electrical energy storage systems. In: Proceedings of the Design, Automation & Test in Europe Conference & Exhibition (DATE), pp 1060–1065
11. Xie Q, Wang Y, Kim Y, Shin D, Chang N, Pedram M (2012b) Charge replacement in hybrid electrical energy storage systems. In: Proceedings of the Asia and South Pacific Design Automation Conference (ASP-DAC), pp 627–632
12. Xie Q, Kim Y, Wang Y, Kim J, Chang N, Pedram M (2013a) Principles and efficient implementation of charge replacement in hybrid electrical energy storage systems. Power Electronics, IEEE Transactions on PP(99):1–1, DOI 10.1109/TPEL.2013.2295601
13. Xie Q, Wang Y, Kim Y, Pedram M, Chang N (2013b) Charge allocation in hybrid electrical energy storage systems. IEEE Transactions on Computer-Aided Design of Integrated Circuits and Systems 32(7):1003–1016, DOI 10.1109/TCAD.2013.2250583
14. Xie Q, Zhu D, Wang Y, Kim Y, Chang N, Pedram M (2013c) An efficient scheduling algorithm for multiple charge migration tasks in hybrid electrical energy storage systems. In: Proceedings of the Asia and South Pacific Design Automation Conference (ASP-DAC) (to appear)
15. Zhang Y, Jiang Z, Yu X (2008) Control strategies for battery/supercapacitor hybrid energy storage systems. In: Proceedings of the IEEE Energy 2030 Conference, pp 1–6

Chapter 4
Architectures for Energy Efficiency

In this chapter, we propose two novel architectures for improving energy efficiency. The first architecture that we introduce is a reconfigurable EES array architecture. It enables dynamic changes of series and parallel configurations of an EES array so that power conversion efficiency can be maximized. The second architecture is a networked CTI architecture. This CTI architecture provides a high scalability than single shared bus CTI architecture; it can accommodate an increased number of nodes without a significant energy efficiency degradation. The content of this chapter is in part based on [7, 8].

4.1 Modeling Power Conversion Efficiency

A *power converter* is to deliver regulated voltage or current to the energy storage at a desired level regardless of variation in the input power source. The power converter is an essential component to resolve voltage or current mismatch between the input and output, and to provide controlled power delivery. A general schematic of a buck-boost switching power converter is shown in Fig. 4.1. Depending on the relation between V_{in} and V_{out}, a power converter has two working modes: buck (step-down) mode and boost (step-up) mode. As the names imply, power converters operate in the buck mode if $V_{in} > V_{out}$, and otherwise in the boost mode.

An ideal power converter delivers the entire power from the source to the load without any loss, but the power conversion involves non-zero amount of power loss in practice. The power converter efficiency η_c is defined as

$$\eta_c = \frac{P_{out}}{P_{in}} = \frac{P_{in} - P_c}{P_{in}} = \frac{V_{in} \cdot I_{in} - P_c}{V_{in} \cdot I_{in}}, \qquad (4.1)$$

Y. Kim, N. Chang, *Design and Management of Energy-Efficient Hybrid Electrical Energy Storage Systems*, DOI 10.1007/978-3-319-07281-4_4

where P_{in} and P_{out} is input and output power of the power converter, respectively, and P_c is the power dissipation of the power converter. Therefore,

$$P_{out} = P_{in} - P_c = \eta_c \cdot P_{in}. \tag{4.2}$$

A pulse-width modulation (PWM) switching power converter is a common type of switching power converter. The power loss of the PWM switching power converter consists of three components: conduction loss P_{cdct}, switching loss P_{sw} and controller loss P_{ctrl} [4]. That is,

$$P_c = P_{cdct} + P_{sw} + P_{ctrl}. \tag{4.3}$$

Those power loss components are strongly dependent on the input voltage V_{in}, output voltage V_{out}, output current I_{out}, and the circuit component properties.

In the buck mode, the power loss components is presented as

$$\begin{aligned} P_{cdct} &= {I_{out}}^2 \cdot (R_L + D \cdot R_{sw1} + (1 - D) \cdot R_{sw2} + R_{sw4}) \\ &\quad + \frac{(\Delta I)^2}{12} \cdot (R_L + D \cdot R_{sw1} + (1 - D) \cdot R_{sw2} + R_{sw4} + R_C), \\ P_{sw} &= V_{in} \cdot f_s \cdot (Q_{sw1} + Q_{sw2}), \\ P_{ctrl} &= V_{in} \cdot I_{ctrl}, \end{aligned} \tag{4.4}$$

where $D = \dfrac{V_{out}}{V_{in}}$ is the PWM duty ratio and $\Delta I = \dfrac{V_{out} \cdot (1 - D)}{L_f \cdot f_s}$ is the maximum current ripple; f_s is the switching frequency; I_{ctrl} is the current flowing into the controller; R_L and R_C are the equivalent series resistance (ESR) of inductor L and capacitor C, respectively; $R_{sw1,\ldots,4}$ and $Q_{sw1,\ldots,4}$ are the turn-on resistances and gate charges of the four switches in Fig. 4.1, respectively.

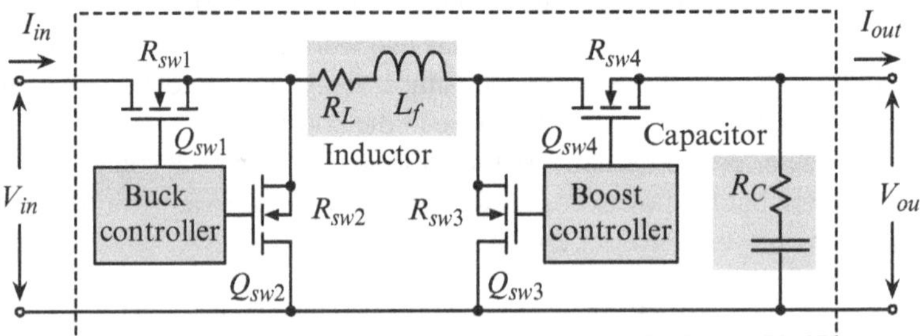

Fig. 4.1 Buck-boost switching power converter model

In the boost mode, the power loss components is presented as

$$
\begin{aligned}
P_{cdct} =& \left(\frac{I_{out}}{D}\right)^2 \cdot (R_L + (1-D) \cdot R_{sw3} + D \cdot R_{sw4} + R_{sw1} + D \cdot (1-D) \cdot R_C) \\
&+ \frac{(\Delta I)^2}{12} \cdot (R_L + (1-D) \cdot R_{sw3} + D \cdot R_{sw4} + R_{sw1} + D \cdot R_C), \\
P_{sw} =& V_{out} \cdot f_s \cdot (Q_{sw3} + Q_{sw4}), \\
P_{ctrl} =& V_{in} \cdot I_{ctrl},
\end{aligned} \tag{4.5}
$$

where $D = \frac{V_{in}}{V_{out}}$ and $\Delta I = \frac{V_{in} \cdot (1-D)}{L_f \cdot f_s}$.

The gate width of the MOSFET switches W_{sw} is the determining factor of R_{sw} and Q_{sw}. As W_{sw} gets smaller, R_{sw} increases and Q_{sw} decreases [9, 16]. More specifically,

$$R_{sw} = \frac{W_0}{W_{sw}} \cdot R_0, \tag{4.6}$$

$$Q_{sw} = \frac{W_{sw}}{W_0} \cdot Q_0, \tag{4.7}$$

where R_0 and Q_0 are the turn-on resistance and gate charge, respectively, of a MOSFET switch with a gate width of W_0.

Figure 4.2 shows the efficiency variation of the LTM4609 buck-boost converter from Linear Technology [10]. It shows a wide range of variation depending on V_{in}, V_{out}, and I_{out}. This wide variation may cause that the optimal operating point of the PV module does not match to the system-level optimal operating point.

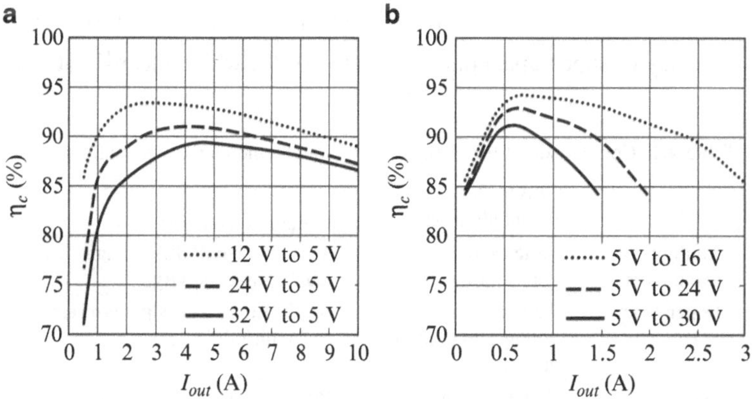

Fig. 4.2 Power conversion efficiency of the Linear Technology LTM4609 buck-boost converter [10]. (**a**) Buck mode. (**b**) Boost mode

4.2 Reconfigurable EES Array Architecture

Memory hierarchy and organization has a significant impact on the computer system performance and power consumption. Objectives of memory system design include reducing access time, power consumption, cache miss rate, and so on. However, a memory design optimized for one application may not be optimal for all applications. Therefore, it is beneficial to adaptively change the organization of the memory design dynamically depending on the access pattern of the currently running applications [1, 21]. This is called memory reconfiguration.

A similar problem exists in the HEES system. One configuration of an EES bank may not be optimal always depending on the current cell voltage, CTI voltage, and amount of current. Therefore, we introduce EES bank reconfiguration that adaptively changes the series and parallel connection of the EES array. It aims to maximize the energy efficiency by reducing power loss in the power converter and EES element and maximize capacity utilization. This comparison is summarized in Table 4.1

4.2.1 *Cycle Efficiency and Capacity Utilization of EES Bank*

Cost factors of HEES systems fall into two categories: operational cost and capital cost [14]. The operational cost is mainly the electricity cost, and thus it is directly related to the efficient use of energy. The capital cost includes expenses for purchasing and disposal of the EES elements, and therefore fully utilizing the EES bank capacity is a key for reducing the capital cost. Cycle efficiency and capacity utilization of EES banks are the major factors that motivate dynamic reconfiguration of EES banks for reducing the operational cost and capital cost of HEES systems.

The cycle efficiency is 'round-trip' energy efficiency generally defined as $\eta_{cyc} = \frac{E_{out}}{E_{in}}$ where E_{in} and E_{out} denote energy input and energy output, respectively. The cycle efficiency of supercapacitors is close to 100 %, which means that almost all

Table 4.1 Comparison of EES bank reconfiguration and memory reconfiguration

	EES bank reconfiguration	Memory reconfiguration
Configurable parameters	Series and parallel connections	Memory size, (DRAM) rank size, (cache) associativity
Dependency	Cell voltage, CTI voltage, amount of current	Applications access pattern (e.g., cache hit/miss rates)
Objectives	Reducing power loss in power converter and EES elements	Reducing access time, power consumption, (cache) miss rate

the energy consumed to charge a supercapacitor can be retrieved in the following discharging process. On the other hand, the cycle efficiency of batteries ranges 60–90 % depending on the chemistry used for the electrodes even under the optimal charge and discharge condition.

The cycle efficiency generally has been considered as a natural characteristics of an EES element [3]. However, the cycle efficiency is closely related to the charge and discharge rates, i.e., the magnitude of charge and discharge current with respect to the rated capacity of the storage element. The rate capacity effect of batteries results a low cycle efficiency for a high-current charge and discharge. In practice, from the system-level point of view, one should not disregard the power conversion process and its power loss when considering the cycle efficiency of an EES bank. In fact, the cycle efficiency is significantly affected by the power conversion efficiency, which is also a function of the charge and discharge rates as discussed in Sect. 4.1.

Therefore, it is beneficial to define constant-power charging efficiency η_c and constant-power discharging efficiency η_d for the cases that a storage element is charged and discharged at a constant CTI power and a constant CTI voltage:

$$\eta_c(P_{cti,c}, V_{cti}) = \frac{\max(E_{bank})}{P_{cti,c} \cdot t_c(P_{cti,c}, V_{cti})}, \tag{4.8}$$

$$\eta_d(P_{cti,d}, V_{cti}) = \frac{P_{cti,d} \cdot t_d(P_{cti,d}, V_{cti})}{\max(E_{bank})}, \tag{4.9}$$

where $t_c(P_{cti,c}, V_{cti})$ is the charging time, $t_d(P_{cti,d}, V_{cti})$ is the discharging time, $P_{cti,c}$ is the CTI power when charging, $P_{cti,d}$ is the CTI power when discharging, and V_{cti} is the CTI voltage. As a result, the *constant-power cycle efficiency* η_{cyc} when $P_{cti,c} = P_{cti,d} = P_{cti,cyc}$ is defined as

$$\eta_{cyc}(P_{cti,cyc}, V_{cti}) = \eta_c(P_{cti,cyc}, V_{cti}) \cdot \eta_d(P_{cti,cyc}, V_{cti}). \tag{4.10}$$

We define *capacity utilization* as one of the important performance metrics of an EES bank. A bank voltage cannot be arbitrarily low because the power converter cannot operate below a certain voltage [5], which we define as $V_{bank,min}$. The capacity utilization ρ is defined as the ratio between the usable energy capacity and the original energy capacity of the EES bank. The capacity utilization is equivalent to the ratio between the extracted energy and the stored energy in a fully charged bank. That is,

$$\rho = 1 - \frac{E_{bank,remain}}{E_{bank,lim}}, \tag{4.11}$$

where

$$E_{bank,remain} = \frac{1}{2} \cdot C_{bank} \cdot {V_{bank,min}}^2 \tag{4.12}$$

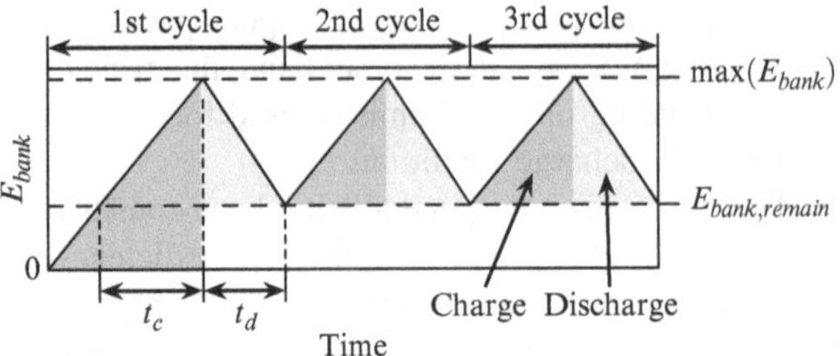

Fig. 4.3 Cycle efficiency and capacity utilization in repeated charge-discharge cycles

is the remaining energy when the power converter can no longer extract energy, i.e., loss in the capacity. The capacity utilization of storage elements is smaller than 100 % because the power converter requires the minimum bank voltage, $V_{bank,min}$, which is higher than 0 V.

Figure 4.3 illustrates the effect of the cycle efficiency and capacity utilization. The operational cost of HEES systems is affected mainly by the cycle efficiency. If the cycle efficiency is poor, we have to expense more for storing and retrieving the same amount of energy. As shown in Fig. 4.3, the very first cycle requires additional energy of $E_{bank,remain}$ to increase the bank voltage from 0 V to $V_{bank,min}$. The operational cost for this additional energy is dependent on the capacity utilization, but this effect may be neglected for repeated cycles in long term. The capital cost is affected by the effective energy capacity of the EES banks. If the capacity utilization is 80 %, we lose 20 % of the expenses for the storage elements because this portion does not contribute to the energy capacity.

4.2.2 General Bank Reconfiguration Architecture

We introduce the *general balanced reconfiguration architecture (GBRA)* for the EES bank reconfiguration [7]. We define a balanced reconfiguration to satisfy the condition whereby all energy storage cells in a given EES bank have identical SoC and terminal voltages at all times i.e., they are balanced at all times, given that the cells are healthy and identical. Unless active charge balancing circuits are used (which is not the case here), cell balancing can be achieved by regular arrangement of cells. We call such arrangements balanced configurations. The proposed architecture is 'general' in the sense that it can produce every balanced configuration that is possible with a given number of cells.

Let N denote the number of available cells. The N cells can be organized in various balanced configurations and the number of possible configurations equals the number of bi-factor decompositions of the natural number N (including $1 \times N$ and $N \times 1$). We define a configuration $\mathcal{C}(n, m)$ to be a configuration that has n cells in series and m cells in parallel. For example, the number of balanced configurations of a 10-cell bank (N is 10) is four: $\mathcal{C}(1, 10), \mathcal{C}(2, 5), \mathcal{C}(5, 2)$, and $\mathcal{C}(10, 1)$. Although $\mathcal{C}(3, 3)$, which is composed of nine cells, is also possible with 10 cells, we do not consider such a case as a balanced configuration because it leaves one cell imbalanced.

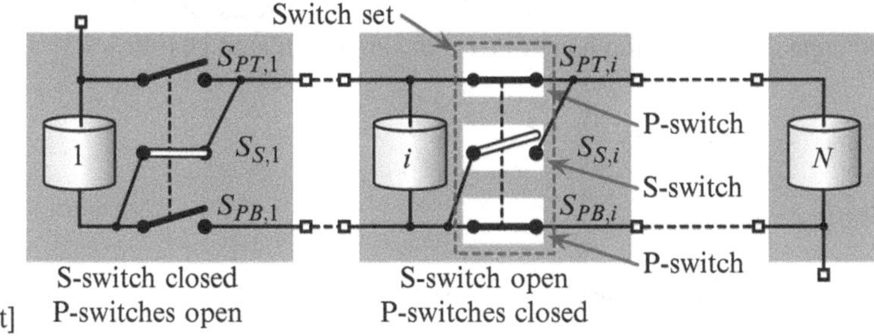

Fig. 4.4 GBRA of an N-cell EES bank

Figure 4.4 shows the proposed GBRA of a bank composed of N cells. Each of $N-1$ cells has three switches: one series switch (S-switch) and two parallel switches (P-switches) except for the last one. The P-switches connect cells in parallel into n-parallel sub-banks, whereas the S-switches connect those m sub-banks in series. A sub-bank is a set of cells connected only in parallel. For the i-th cell, its S-switch is denoted by $S_{S,i}$, while its two P-switches are denoted by $S_{PT,i}$ and $S_{PB,i}$, one on the top and the other in the bottom, respectively. We group the three switches of one cell as a switch set, which gives rise $N-1$ switch sets in the bank. For each cell, $S_{PT,i}$ and $S_{PB,i}$ are closed exactly if $S_{S,i}$ is open. There are no cases where one of $S_{PT,i}$ and $S_{PB,i}$ switches is open while the other is closed.

More formally, for $i = 1, 2, \ldots, N-1$,

$$x_{P,i} + x_{S,i} = 1, \tag{4.13}$$

where

$$x_{S,i} = \begin{cases} 0 & \text{if } S_{S,i} \text{ is open,} \\ 1 & \text{if } S_{S,i} \text{ is closed,} \end{cases} \tag{4.14}$$

$$x_{P,i} = \begin{cases} 0 & \text{if } S_{PT,i} \text{ and } S_{PB,i} \text{ are open,} \\ 1 & \text{if } S_{PT,i} \text{ and } S_{PB,i} \text{ are closed.} \end{cases} \tag{4.15}$$

Otherwise, the bank malfunctions; more precisely, the i-th cell is disconnected from the $(i+1)$-th cell if $x_{S,i} = x_{P,i} = 0$, or the i-th supercapacitor is short-circuited if $x_{S,i} = x_{P,1} = 1$.

A balanced configuration of GBRA is obtained by switching operations which obeys the following rule; in the m-by-n balanced configuration $C(n, m)$,

$$x_{S,i} = \begin{cases} 1 & \text{if } i = n \cdot k \text{ where } k = 1, 2, \ldots, m-1, \\ 0 & \text{otherwise,} \end{cases} \tag{4.16}$$

$$x_{P,i} = 1 - x_{S,i}. \tag{4.17}$$

The total capacitance C_{bank}, voltage V_{bank}, internal resistance R_{bank}, and energy storage capacity E_{bank} of a bank of $\mathcal{C}(n, m)$ are calculated as follows:

$$C_{bank} = \frac{n}{m} \cdot C_{cell} = \frac{N}{m^2} \cdot C_{cell}, \tag{4.18}$$

$$V_{bank} = m \cdot V_{cell}, \tag{4.19}$$

$$E_{bank} = \frac{1}{2} \cdot C_{bank} \cdot {V_{bank}}^2 = N \cdot E_{cell}, \tag{4.20}$$

$$R_{bank} = \left(\frac{2}{3} \cdot n - 1 + \frac{1}{3 \cdot n}\right) \cdot m \cdot R_p + \frac{m}{n} \cdot R_c + (m-1) \cdot R_s, \tag{4.21}$$

where C_{cell}, V_{cell}, and E_{cell} denote the capacitance, voltage, and energy capacity of each cell, respectively; R_s and R_p denote the on-resistance of an S-switch and a P-switch, respectively; and R_c denotes the ESR of each cell. We assume that the charge or discharge current is equally distributed to every cell in a sub-bank when we derive (4.21). For a fixed N, the bank total capacitance C_{bank} is inversely proportional to m^2 whereas the bank terminal voltage V_{bank} is proportional to m. The total energy remains the same regardless of the configuration.

Each cell has its voltage limit $V_{cell,lim}$ that should not be exceeded, and corresponding energy capacity limit $E_{cell,lim}$. The voltage limit $V_{bank,lim}$ and energy capacity limit $E_{bank,lim}$ of a bank are defined similar to (4.19) and (4.20):

$$V_{bank,lim} = m \cdot V_{cell,lim}, \tag{4.22}$$

$$E_{bank,lim} = N \cdot E_{cell,lim}. \tag{4.23}$$

Figure 4.5 is an example of reconfiguration of a four-cell bank ($N = 4$) with the GBRA. With four cells, three balanced configurations are possible. One of them, for example, is $\mathcal{C}(2, 2)$ which consists of two sub-banks connected in parallel with P-switches, and each sub-bank composed of two cells connected in series with S-switches.

Figure 4.6 shows the switch operations for each configuration when $N = 60$. Each row represents the configuration $\mathcal{C}(n, n)$, and each square represents which of the S-switch and P-switches are closed in the configuration. One can notice from the figure that each switch set has a different probability for closing the S-switch or P-switches. For example, $S_{S,30}$ is more likely to be closed than other S-switches in many configurations. On the other hand, switch sets that are annotated with dotted boxes always close P-switches except only for one configuration $\mathcal{C}(60, 1)$. To generalize, $x_{S,i} = 1$ in $\mathcal{C}(n, m)$ exactly if m is a common divisor of i and N, otherwise, $x_{P,i} = 1$ as shown in (4.16).

This observation provides the intuition for an optimization method to reduce the number of switches. A switch set can be removed if the switches in the set do not change their states, i.e., they remain always open or always closed. An always-open switch may be removed from the circuit, while an always-closed switch may be

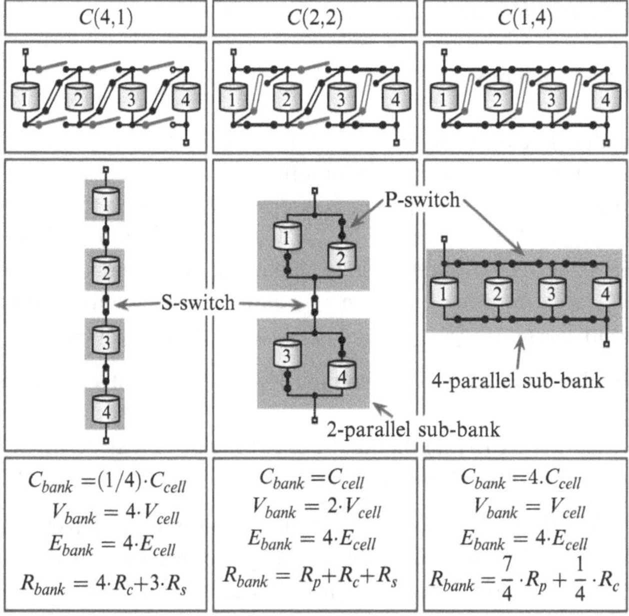

Fig. 4.5 Reconfiguration examples of a four-cell EES bank ($N = 4$)

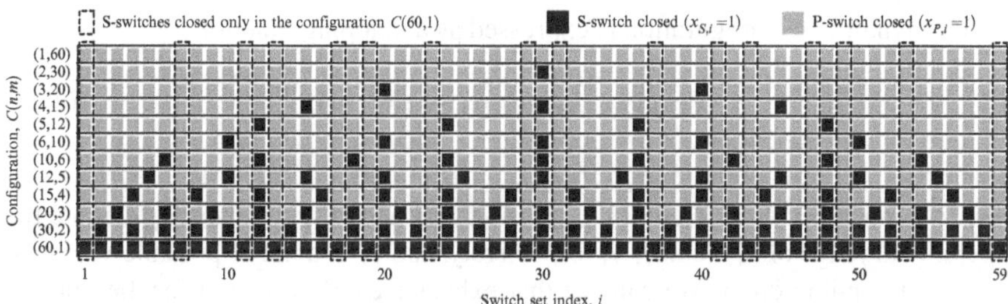

Fig. 4.6 Operations of S-switches and P-switches of a 60-cell bank in different configurations

replaced by a wire. Eliminating unnecessary switch sets not only reduces the overall switch implementation cost, but also reduces the bank internal resistance and in turn, lowers the IR loss.

We can reduce the number of switches by restricting possible configurations. More precisely, $S_{S,i}$ may be short-circuited, and $S_{PT,i}$ and $S_{PB,i}$ may be open if we use only configurations where n and i are coprime. Conversely, $S_{S,i}$ may be open, and $S_{PT,i}$ and $S_{PB,i}$ may be short-circuited if we never use configurations where n and i are coprime. In the previous example, out of the 59 switch sets, 16 switch sets that are annotated with the dotted boxes can be removed if $C(60, 1)$ is not used. However, we do not consider configuration selection, and assume all the configurations are possible with all the switch sets present.

4.2.3 Dynamic Reconfiguration Algorithm

4.2.3.1 Cycle Efficiency

The primary objective of the dynamic EES bank reconfiguration is reducing energy loss by improving the power conversion efficiency. As discussed in Sect. 4.1, the power conversion efficiency of power converters depends on the input and output voltage and current values, that is V_{bank}, V_{cti}, I_{bank}, and I_{cti}. The purpose of dynamic EES bank reconfiguration is to maximize conversion efficiency η_{conv} by controlling V_{bank} at run time for given V_{cti} and I_{cti}. Meanwhile, the bank voltage should be within a range of $[V_{bank,min}, V_{bank,max}]$ as the power converter requires.

- Given: Number of cells N, CTI voltage V_{cti}, CTI current I_{cti}, and cell voltage V_{cell}.
- Find: Configuration $\mathcal{C}(n, m)$ that minimizes the power loss of the EES bank.
- Subject to: Bank voltage limitation: $V_{bank,min} \leq V_{bank} \leq V_{bank,max}$.

The power loss of the EES bank has two components: power conversion loss which is discussed in Sect. 4.1, and IR loss induced by the internal resistance of the EES bank. Minimizing the power converter loss has different implications for charging and discharging: (i) for charging, it means maximizing energy transferred from the CTI to the bank, and (ii) for discharging, it means maximizing energy transferred from the bank to the CTI.

The dynamic reconfiguration is expressed as a mapping function

$$f : (V_{cti}, I_{cti}, V_{cell}) \rightarrow \mathcal{C}(n, m), \tag{4.24}$$

where $V_{cti,min} \leq V_{cti} \leq V_{cti,max}$, $I_{cti,min} \leq I_{cti} \leq I_{cti,max}$, $0 \leq V_{cell} \leq V_{cell,lim}$, $n \in \hat{N}$, and $m = N/n$. Here, $[V_{cti,min}, V_{cti,max}]$ and $[I_{cti,min}, I_{cti,max}]$ denote the operational range of V_{cti} and I_{cti}, respectively, and $\hat{N}$ is a list of possible values of n in an ascending order. We can see that exhaustive online search for the optimal n and m among numerous configurations is not practical. Therefore, we propose a two-phase reconfiguration method, which consists of an offline phase and an online phase. In the offline phase, we analyze the power converter efficiency η_{conv} and develop a function $f_{offline}$ to find the optimal bank voltage $V_{bank,opt}$ for given V_{cti} and I_{cti}. Next, in the online phase, we use a function f_{online} to find the optimal configuration $\mathcal{C}(n, m)$ that minimized the power loss for given V_{cell}. That is,

$$f_{offline} : (V_{cti}, I_{cti}) \rightarrow V_{bank,opt}, \tag{4.25}$$

$$f_{online} : (V_{cti}, I_{cti}, V_{bank,opt}, V_{cell}) \rightarrow \mathcal{C}(n, m). \tag{4.26}$$

It is not feasible to analytically find the optimal operating conditions that maximizes the power conversion efficiency η_{conv}. Therefore, it is reasonable to implement the offline function $f_{offline}$ as a lookup table since it is only two-dimensional and both of V_{cti} and I_{cti} have a limited range in practice because of the minimum and maximum ratings of peripheral circuitry. We build the lookup table by evaluating the conversion efficiency and finding the optimal condition. The lookup table is indexed with V_{cti} and I_{cti}, where each entry is the optimal bank voltage $V_{bank,opt}$ that maximizes η_{conv} for given V_{cti} and I_{cti}. Two lookup tables are built for charging and discharging in the same manner. The online algorithm can exploit these lookup tables and easily obtain $V_{bank,opt}$ at run time which greatly reduces the computation overhead.

The online function f_{online} is described in Algorithm 1. First, the optimal bank voltage is derived from $f_{offline}$ mapping function for current V_{cti} and I_{cti} (List 1). Since the lookup table is defined for discrete intervals, a two-dimensional interpolation may be used for intermediate values. Next, the ideal series-connection number n_{ideal} is derived (List 2). We redefine a possible set of configurations M' for the given condition, by excluding configurations that are not allowed due to bank voltage limitation (List 3). If n_{ideal} implies a possible configuration (List 4), this is the optimal value for m. However, it is possible that n_{ideal} represents not a feasible configuration. If n_{ideal} is out of boundary of possible configurations, we set n_{opt} to the minimum or maximum (Lists 6 and 8). Otherwise, we find a consecutive n_i and n_{i+1} in $\hat{N}'$ that are near n_{ideal} (List 11). Between two configurations, we select the one whose sum of the power converter loss and IR loss due to the bank internal resistance is smaller (List 12). Finally, n_{opt} is derived (List 13), and the optimal configuration is returned. This algorithm has $O(\log|\hat{N}|)$ time complexity if $\hat{N}$ is in an ascending order, because finding elements in Lists 3 and 11 can be done with a binary search. Other operations are done in constant time; the lookup table indexing and the interpolating are done in constant time, and efficiency evaluation in List 12 is done only for two configurations regardless of the size of $\hat{N}$.

In a discrete-time reconfiguration scheme, bank reconfiguration is performed every decision epoch, assuming that the voltage and current condition is not significantly changed within a time interval. On the other hand, in a continuous-time reconfiguration scheme, we determine whether if a reconfiguration is needed when the voltage or current condition significantly changes.

Figure 4.7 shows two configuration transitions of a 120-cell bank when discharging. The figure shows transitions from $\mathcal{C}(24,5)$ to $\mathcal{C}(30,4)$, and from $\mathcal{C}(30,4)$ to $\mathcal{C}(40,3)$. The transitions occur at the points where the power loss (sum of power converter loss and internal resistance IR loss) of two consecutive configurations cross. This is different from a previous work [5] that the configuration transitions is triggered by the bank voltage variation constraint. Although limiting the bank voltage variation may improve the power conversion efficiency if the voltage range is chosen elaborately, but there is no explicit clue for setting the voltage range. Furthermore, the current which also affects the conversion efficiency is not considered for reconfiguration in the previous work. The proposed method exhibits a better efficiency since it considers the conversion efficiency for the reconfiguration taking the voltage and current into account.

Algorithm 1: f_{online}: Online optimal configuration determination

Input: V_{cti}: CTI voltage, I_{cti}: CTI current, and V_{cell}: cell voltage
Output: Optimal configuration (n_{opt}, m_{opt})
Global: N: number of cells, $\hat{N}$: list of possible values of m in an ascending order, $f_{offline}$: optimal bank voltage mapping function, P_c: power converter loss model, P_{int}: bank internal resistance IR loss model, $[V_{bank,min}, V_{bank,max}]$: range of V_{bank}

$V_{bank,opt} = f_{offline}(V_{cti}, I_{cti})$
$n_{ideal} = \frac{V_{bank,opt}}{V_{cell}}$
$\hat{N}' = \left\{ m \in \hat{N} \middle| V_{bank,min} \leq m \cdot V_{cell} \leq V_{bank,max} \right\}$
if $n_{ideal} \in \hat{N}'$ **then**
  $n_{opt} = n_{ideal}$
else if $n_{ideal} \leq \min(\hat{N}')$ **then**
  $n_{opt} = \min(\hat{N}')$
else if $n_{ideal} \geq \max(\hat{N}')$ **then**
  $n_{opt} = \max(\hat{N}')$
else
  Find i such that $n_i \leq n_{ideal} \leq n_{i+1}$, where $n_i, n_{i+1} \in \hat{N}'$
  $n_{opt} = \arg\min_n (P_c(V_{cti}, I_{cti}, n \cdot V_{cell}) + P_{int}(n, I_{bank}))$ for $n \in \{n_i, n_{i+1}\}$
$m_{opt} = \frac{N}{n_{opt}}$
return (n_{opt}, m_{opt})

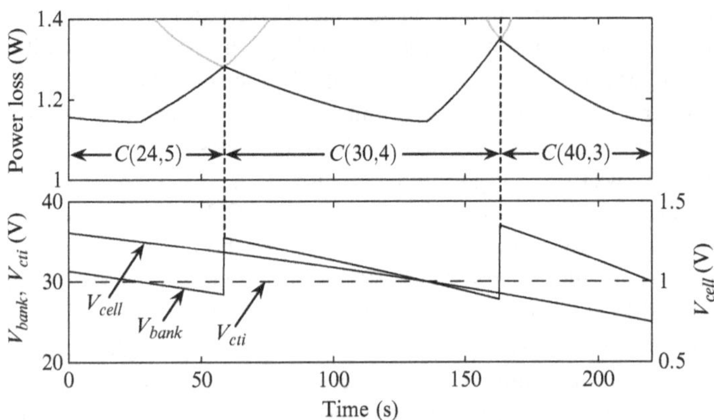

Fig. 4.7 Two configuration transitions before and after the configuration $C(30, 4)$ when discharging a 120-cell bank. CTI voltage (V_{cti}) is 30 V and CTI current (I_{cti}) is -1 A

4.2.3.2 Capacity Utilization

We analyze the capacity utilization improvement by the proposed reconfiguration method. From (4.12) and (4.18), the remaining energy $E_{bank,remain}$ of a configuration $\mathcal{C}(n,m)$ when $V_{bank} = V_{bank,min}$ is

$$E_{bank,remain} = \frac{1}{2} \cdot \frac{n}{m} \cdot C_{cell} \cdot V_{bank,min}{}^2. \tag{4.27}$$

From (4.11), (4.12), (4.23), and that $N = m \cdot n$,

$$\rho = 1 - \left(\frac{\min\left(V_{bank,min}, m \cdot V_{cell,lim}\right)}{m \cdot V_{cell,lim}} \right)^2. \tag{4.28}$$

This implies that a reconfiguration in a way that increases m improves the capacity utilization. Figure 4.8 is an example that graphically shows how the capacity utilization is improved by the reconfiguration. Here, $N = 2$, and two configurations $\mathcal{C}(2,1)$ and $\mathcal{C}(1,2)$ are available. The horizontal and vertical lengths of the box are proportional to the capacitance and square of the voltage, respectively, and so the area is proportional to the energy. Reconfiguration changes the way to store the same amount of energy; either in higher voltage and smaller capacitance (switching to a more series configuration), or in lower voltage and larger capacitance (switching to a more parallel configuration). Therefore, when the bank is deeply depleted and V_{bank} is near $V_{bank,min}$, we can maximize the capacity utilization by reconfiguring the bank to a configuration with the maximum m, that is, N.

4.2.4 *Cycle Efficiency and Capacity Utilization Improvement*

In the experiments, we demonstrate that the proposed EES bank reconfiguration method improves the cycle efficiency and capacity utilization of an EES bank. Throughout this section, we use a supercapacitor bank consisting of capacitors with $C_{cell} = 100$ F and $V_{cell,lim} = 2.5$ V.

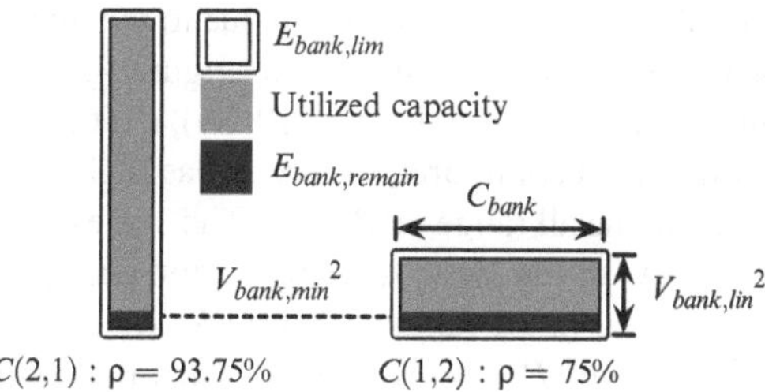

Fig. 4.8 Capacity utilization (ρ) of two configurations of a two-cell bank when $V_{bank,min} = \frac{1}{2} \cdot V_{cell,lim}$. Note that the vertical length is proportional to the square of voltage so that the area is proportional to the energy capacity

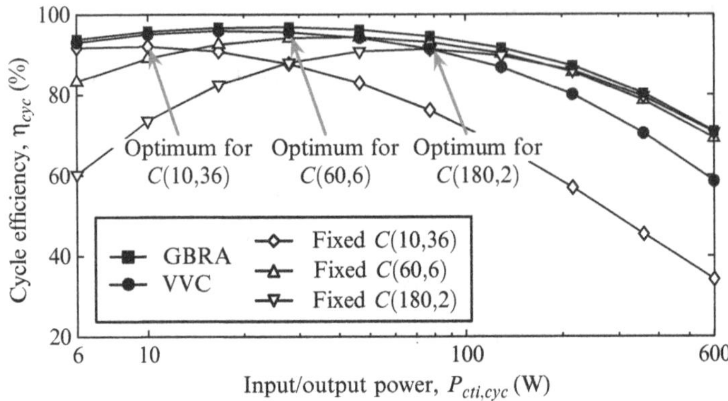

Fig. 4.9 Constant-power cycle efficiency comparison for different input/output power values of a 360-cell bank. CTI voltage (V_{cti}) is 30 V

First, we demonstrate the energy efficiency improvement of the proposed EES bank reconfiguration method (GBRA) compared with two baselines: (i) fixed EES bank configurations (Fixed), and (ii) voltage variation-constraint (VVC) reconfiguration. The VVC reconfiguration method limits the bank voltage variation by switching the configuration when the bank voltage goes out of the given voltage range. We set the range of bank voltage to $\frac{1}{2} \cdot V_{cti} \leq V_{bank} \leq \frac{3}{2} \cdot V_{cti}$ for the VVC reconfiguration in the experiment. We assume a low minimum bank voltage constraint of $V_{bank,min} = 1.25$ V to minimize the effect of the capacity utilization limit.

We first demonstrate that the constant-power cycle efficiency η_{cyc} discussed in Sect. 4.2.1, is improved by the proposed reconfiguration method. Figure 4.9 shows the constant-power cycle efficiency of a 360-cell bank according to $P_{cti,cyc}$ ranging from 6 to 600 W, when $V_{cti} = 30$ V.

We can see that the proposed GBRA reconfiguration exhibits the best cycle efficiency for the all range by the timely efficiency-aware reconfiguration. On the other hand, the cycle efficiency of the VVC reconfiguration is lower than that of the GBRA reconfiguration, especially when the input/output power $P_{cti,cyc}$ is large. This is because the efficiency degradation due to the input and output voltage difference is escalated as the current increases. Figure 4.9 also shows the cycle efficiency of the three fixed configurations: $\mathcal{C}(10, 36)$, $\mathcal{C}(60, 6)$, and $\mathcal{C}(180, 2)$. The cycle efficiencies of the fixed configurations are not as high as that of the proposed GBRA reconfiguration in the all range of $P_{cti,cyc}$. The cycle efficiency improvement is up to 21 % compared with the VVC reconfiguration and up to 108 % compared with the fixed configurations in the range of 6–600 W. We can see from Fig. 4.9 that the cycle efficiency for larger $P_{cti,cyc}$ is optimal when the configuration has more cells in series. This is because a low bank voltage results in an excessively large

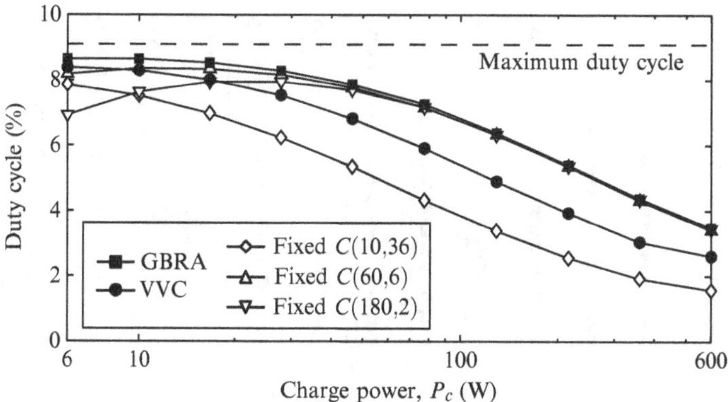

Fig. 4.10 Duty cycle for a high-current pulsed load of a 360-cell bank when discharge power (P_d) is 10 times higher than charge power (P_c). CTI voltage (V_{cti}) is 30 V

current for larger $P_{cti,cyc}$, and induces a large P_{cdct} loss. In contrast, the P_{sw} loss becomes dominant for the high bank voltage cases. This clearly shows that finding the optimal configuration is not straightforward.

Next, we demonstrate the energy efficiency improvement with a varying power input and output, which is more realistic for practical HEES systems. We charge the bank with a low input power until it is fully charged, and then discharge it with a high output power until it is fully depleted. The discharge power P_d is 10 times higher than the charge power P_c, and therefore the maximum duty cycle, which is possible only when the cycle efficiency is 100 %, is $\frac{1}{10+1} = 9.1\,\%$. Figure 4.10 shows the duty cycle of a 360-cell bank for P_c ranging from 6 to 600 W, when $V_{cti} = 30$ V. This result also shows that the proposed GBRA reconfiguration exhibits higher energy efficiency even for a realistic high-power pulsed load demand, compared with the VVC reconfiguration and fixed configurations. The duty cycle improvement is by up to 44 % compared with the VVC reconfiguration and by up to 127 % compared with the fixed configurations in the range of 6–600 W input power.

We demonstrate the capacity utilization for different n and $V_{bank,min}$ values in Fig. 4.11. As seen in (4.28), the capacity utilization is dependent on n, but not on m. We can see that the capacity utilization increases as n increases. This result clearly shows the motivation of dynamic reconfiguration to fully utilize the capacity. By dynamically increasing the number of series connections, n, when the bank is almost depleted, we can extract more energy from the bank. For example, when $V_{bank,min} = 10$ V, using a fixed configuration with $n = 10$ results in 84 % of capacity utilization. These imply that 16 % of the capital cost to purchase and dispose the EES elements is wasted without substantial capacity increase. The capital cost loss is reduced to less than 1 % when the EES bank can be reconfigured to $n = 60$.

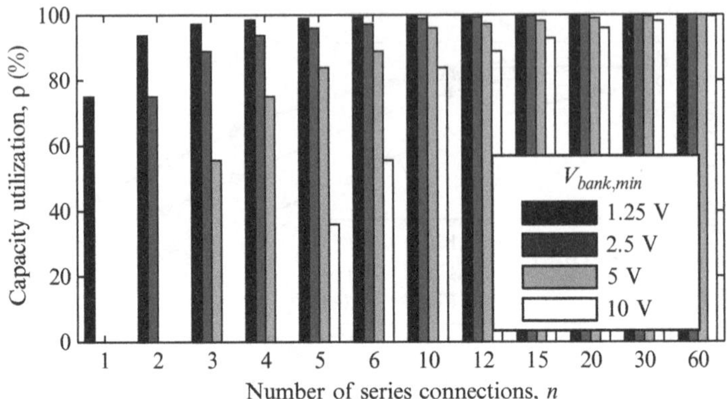

Fig. 4.11 Capacity utilization (ρ) for different number of series connections (n) and the minimum bank voltage requirement ($V_{bank,min}$)

The energy and cost overhead of the proposed reconfiguration architecture is negligible. The energy overhead is caused by the conduction loss in the MOSFET switches due to their non-zero on-resistance. The resistance of power MOSFET switches are typically a few mΩ. The energy loss due to the resistance is only 1.0 % of the total energy in the capacitor array per charge cycle when the charge current is 3 A and a 3 mΩ switch is used. This is negligible when considering the significant energy efficiency improvement. The cost overhead is also insignificant. A switching circuit composed of three pairs of switches and gate drivers are required per cell. The cost of the switching circuit is only a few percent of the total cost including the supercapacitors. For instance, a switch circuit composed of ON Semiconductor NTD4904NT4G and one Maxim MAX15054AUT+T costs only $3. A supercapacitor Maxwell BCAP0650 is as expensive as $43. The PP, which is the time to recover the cost overhead, is dependent to the charge/discharge frequency and electricity cost. The higher charge/discharge frequency or electricity cost is, the shorter PP becomes.

4.3 Networked Charge Transfer Interconnect

4.3.1 Networked Charge Transfer Interconnect Architecture

A HEES system is composed of many EES nodes through a CTI, and so the CTI architecture is an important design factor. It has a significant impact on the charge transfer efficiency, and thus should be carefully designed in order to maximize the benefits of the HEES systems. A system-on-chip is subject to the similar problem of determining a proper interconnect architecture. The interconnect architecture on a system-on-chip affects communication latency, throughput, power consumption,

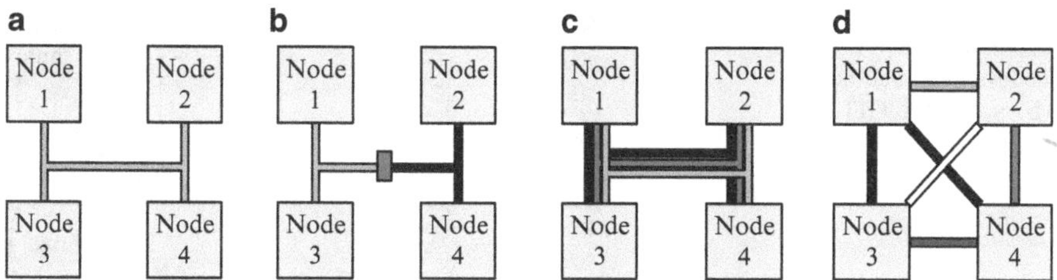

Fig. 4.12 Various interconnect architectures for system-on-chip and HEES systems. (**a**) Shared bus. (**b**) Segmented bus. (**c**) Multiple bus. (**d**) Point-to-point

and so on. For both HEES system and system-on-chip, the interconnect architecture should be selected to considering the scalability. As the number of nodes increases, the interconnect architecture becomes more critical.

Figure 4.12 shows four interconnect architectures. Figure 4.12a is a shared bus interconnect. It is simple to implement, but has a limited scalability. Variances of the shared bus CTI with higher scalability include segmented bus CTI in Fig. 4.12b and multiple bus CTI in Fig. 4.12c. The point-to-point interconnect in Fig. 4.12d provides independent paths between every pair of nodes, but its cost increases exponentially as the number of nodes increases. These architectures are well explored for the system-on-chips, but also applicable for the HEES systems as we shall discuss.

4.3.1.1 Charge Transfer Conflicts

Shared-bus CTI architectures (sometimes called DC bus) are commonly used when the number of EES banks is limited. Recent works on the HEES system management methodologies [7, 18–20] assume a general shared-bus CTI architecture. The shared-bus CTI is analogous to an on-chip shared bus on a system-on-chip and their advantages and disadvantages are similar. Another CTI architecture is a complete point-to-point connection among the nodes [13]. Both the shared-bus and point-to-point connection architectures are feasible as long as the number of EES banks is small, but they certainly lack scalability to accommodate a large-scale HEES system. The other architecture is a customized network architecture for a particular application and operation policy. For example, a supercapacitor buffer efficiently mitigates the rate-capacity effect of a Li-ion battery especially for pulsed load demand [15]. As the control policy is to use the supercapacitor as a buffer of the battery, the path from the battery bank to the load device is not necessary. This architecture is similar to an network-on-chip (NoC) architecture with irregular connectivity which is fully dependent on the application. In short, none of the previously introduced CTI architectures can be used to accommodate a large number of EES banks for general applications.

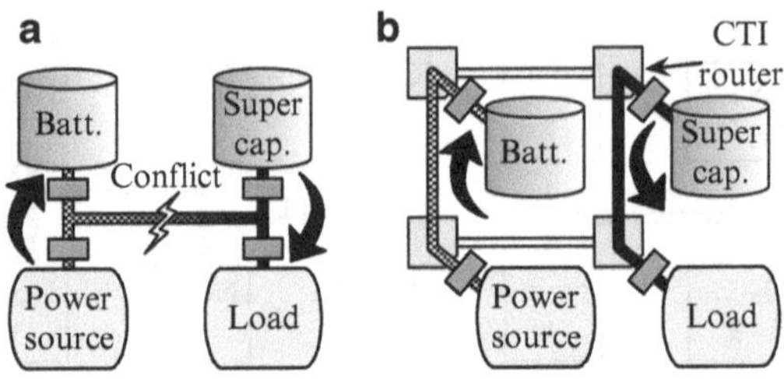

Fig. 4.13 Shared-bus CTI and networked CTI of four nodes. (**a**) Shared-bus CTI. (**b**) Networked CTI

The charge management of a HEES system is achieved by charge allocation, replacement, and migration operations [14]. The operations are basically charge transfers among EES banks using the CTI as charge transfer medium. The previous works on the charge management of HEES [18–20] assumed that the charge transfer path is always available for a given charge transfer task. They focused on maximizing the energy efficiency by setting a proper value for CTI voltage of the charge transfers. However, it is not always true that a charge transfer path is available whenever it is required. Two or more charge transfer tasks can have a conflict by competing for the shared-bus. Figure 4.13a demonstrates an example where the power supply charges the battery bank and the supercapacitor bank supplies power to the load at the same time. Two charge transfer tasks have different optimal CTI voltage values, which maximize the charge transfer energy efficiency of each task, and there is only one CTI link.

We define that two or more charge transfers conflict when they try to occupy the same CTI link and have different optimal CTI voltage values. Such a conflict enforces the charge transfer tasks to use the same CTI voltage, and thus at least one of them has to suffer possibly severe degradation in energy efficiency.

4.3.1.2 Networked CTI Architecture

We introduce a *networked CTI architecture* as shown in Fig. 4.13b to fundamentally solve the charge transfer conflict problem, which ensures scalability to a large number of EES banks [8]. Specifically, we use a mesh interconnect architecture to ensure flexibility and scalability of networked CTI architecture. One important component to realize the networked CTI architecture is the CTI router. We propose a CTI router that connects CTI links, an associated component (i.e., an EES bank, a power source, or a load device), and a power converter. Figure 4.14 shows the detailed architecture of the CTI router. Each CTI router is connected with the adjacent CTI routers through the CTI links. The CTI router consists of reconfigurable interconnects which are denoted as dashed lines in Fig. 4.14. We dynamically connect or disconnect the reconfigurable interconnects inside the router to setup a path from one CTI link to another. The reconfigurable interconnects form a complete graph so that the signal can be routed in any direction. The CTI router

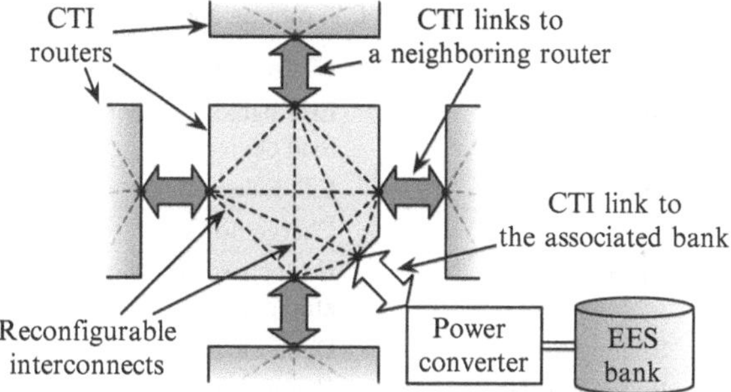

Fig. 4.14 Architecture of a CTI router. An associated EES bank is connected via a power converter. The *arrows* denote the CTI links

in Fig. 4.14 has five CTI links, and thus it has ten interconnects each of which is implemented as a pair of back-to-back MOSFET switches. We adopt the switching power converter efficiency model from [4].

The networked CTI architecture is comparable to a general NoC architecture. As the number of processing elements in an SoC increases, the single-level on-chip bus architecture is no longer able to handle increased data exchanges between the processing elements. Similar to the NoC which requires packet routing, a HEES system with a networked CTI architecture requires routing of the charge transfers. However, CTI routing on a networked CTI is not the same as the conventional NoC packet routing, conventional signal routing for field-programmable gate array (FPGA), nor application-specific integrated circuit (ASIC).

4.3.2 Conventional Routing Problems

The CTI routing problem in a networked CTI has similarity to the conventional FPGA signal routing problem. In the problem of CTI routing, each task competes for routing resources such as converters and CTI links, whereas each signal competes for wires and connection points in FPGA routing. The FPGA routing is a highly complex combinatorial optimization problem, and thus it is usually done by iterative rip-up and reroute of signals. The success of routing is dependent not just on the choice of which nets to reroute, but also on the order in which rerouting is done as shown in traditional rip-up and reroute methods [2,6]. The negotiation-based FPGA router successfully relieves the signal ordering problem and provides a systematic rip-up and reroute capability [12]. This routing algorithm allows initial sharing of the routing resources among signals, but subsequently makes them negotiate for the shared resource with other signals until no resource is shared. The negotiation-based routing algorithm is further enhanced in terms of compilation time by incorporating delay-driven routing [17]. More recent works such as [11] focus on the new architecture or technology scaling, but the core of the routing algorithm is still based on [12]. Table 4.2 briefly compares the PCB/VLSI routing problem and the CTI routing problem for the HEES system.

Table 4.2 Comparison of CTI routing and FPGA/PCB routing

	CTI routing	FPGA/PCB routing
Nodes	EES banks, power sources, load devices	Lookup tables (LUT), ICs
Links	CTI links	Chanel segments, wires
Flows	Charge flows	Signal flows
Objective	High charge transfer efficiency	Low latency
Routing time	Dynamic (runtime)	Static (design time)
Routing output	Charge transfer routing trees	On-/Off-chip interconnect connections
Routing resource sharing	Not allowed	Allowed
Routability	Guaranteed w/resource sharing	Not guaranteed

4.3.3 Routing Problems

We formally describe CTI routing procedures of the networked CTI. We have a set of charge transfer tasks $\tau = \{T_i\}$ to perform. A charge transfer task is defined as a five-tuple $T_i = (\Sigma_i, \Delta_i, \epsilon_i, R_i, D_i)$, where Σ_i is the source node, Δ_i is the destination node, ϵ_i is the amount of energy to be transferred to Δ_i, R_i is the task arrival time, and D_i is the duration. A task T_i and the participating nodes Σ_i and Δ_i are active during the time period of $[R_i, R_i + D_i]$. We assume that all the charge transfer tasks are single-source single-destination without loss of generality.

The CTI routing is a runtime procedure that finds independent routing paths that connects all the nodes in Σ_i and Δ_i for the period of D_i for each $T_i \in \tau^a$, where $\tau^a \subset \tau$ is the set of active charge transfer tasks. The CTI routing procedure assigns the CTI links, which corresponds to edges of E, to each T_i to make the path. The CTI routing procedure involves charge transfer optimization, which is to find the optimal CTI voltage V_{cti}^{opt} of the charge transfer task that maximizes the charge transfer efficiency. The networked CTI should periodically perform the CTI routing to maintain the best CTI configuration at all times for the given τ^a.

However, the number of CTI links is limited, and it may be impossible to assign independent paths to all the tasks if there are too many tasks that should take place simultaneously. The previous work introduces a unique feature of CTI routing such that some tasks may share the CTI link in such cases [8]. This is called task merging. It consolidates the tasks at the expense of charge transfer efficiency degradation because the merged tasks should share a sub-optimal CTI voltage like a shared-bus CTI. Therefore, reducing the number of task merging is the key to maximize the charge transfer efficiency [8].

We present a formal definition of the CTI routing problem in this section. A CTI network is a graph $G = (V, E)$ where V is a set of vertices that corresponds to nodes, and E is a set of edges that corresponds to CTI links between two elements in V. It is an undirected graph as the CTI links are bidirectional electrical conductors. The link between the CTI router and the associated EES element (an EES bank, a power source or a load device) is a dedicated resource, and thus we do not consider this in the routing algorithm.

The CTI routing problem is to find routing paths for a given transfer task set τ, that connects all the nodes in Σ_i and Δ_i for each $T_i \in \tau$. A node of T_i participates in only one charge transfer, and it is either a source or a destination, not both. That is,

$$\bigcup_{T_i \in \tau} (\Sigma_i \cap \Delta_i) = \varnothing \quad \text{and} \quad \bigcap_{T_i \in \tau} (\Sigma_i \cup \Delta_i) = \varnothing. \tag{4.29}$$

As a result of the CTI routing, a disjoint subset of edges in E that forms an acyclic routing tree is assigned to each T_i. We set each CTI router configuration (make connections of the internal interconnects) according to the edges in the routing trees. An individual routed charge transfer is equivalent to a charge transfer on an independent shared-bus CTI. Therefore, it enables us to apply any previous HEES charge management methods that are based on a shared-bus CTI to each routed charge transfer task.

The routing process allocates limited resources to the nets (the set of charge transfer tasks or signals), and each net is allowed to use the resource for a designated period. Routing charge transfer tasks requires iterative execution of two steps; (i) the CTI routing and (ii) charge transfer optimization. The CTI routing operation is to determine a routing path of the charge transfer, and the charge transfer optimization operation is to determine the voltage level of the routing path and the amount of current through the routing path.

The CTI routing problem should tackle limitation in the routing resources (the CTI links) like the conventional FPGA routing problems. Signal routing of FPGA fails if there are unrouted nets which are not routable with remaining routing resources. The workaround is either increasing the resource, i.e., using a larger device or optimizing placement so that the congestion is reduced.

On the other hand, redoing placement is not an option for the CTI routing problem because the nodes are at a fixed location in the HEES system and cannot be moved. Instead, we perform *merging* in order to mitigate the routing congestion. This is a unique feature of the CTI routing for HEES systems. Merging is combining two charge transfer tasks into one to produce a new task set. Two or more migration tasks can be merged and share resources unlike signal routing.

If one task has a longer deadline than the other, the combined task uses the CTI links for whichever the shorter deadline. After the deadline expires, the task with a shorter deadline releases the CTI links and the task with a longer deadline solely occupies the CTI links after rerouting. Merging $T_i = (\Sigma_i, \Delta_i)$ and $T_j = (\Sigma_j, \Delta_j)$ results in a new task $T_{i,j} = (\Sigma_i \cup \Sigma_j, \Delta_i \cup \Delta_j)$. Then T_i and T_j are removed from τ and $T_{i,j}$ is added to τ. After T_i or T_j that has a shorter deadline is finished, the remaining task is added back to τ with the remaining deadline.

Figure 4.15 is an example of the merging to improve routability of three tasks. In Fig. 4.15a, T_2 and T_3 are routed, but T_1 is not routed. There are three possible

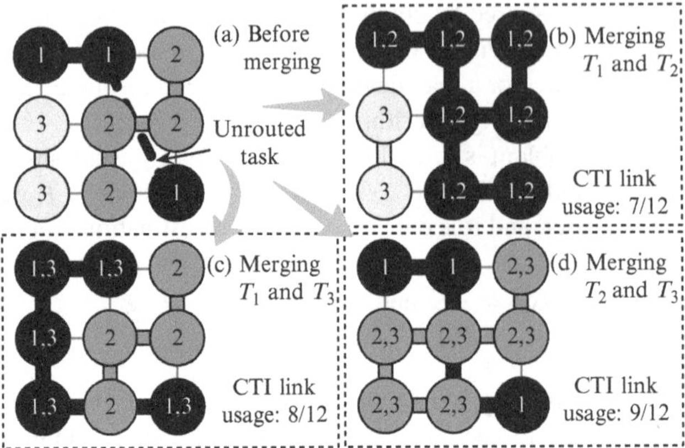

Fig. 4.15 Example routing of three tasks after merging. T_1 is unrouted in (**a**), and routing after three possible merging combinations are presented in (**b**), (**c**), and (**d**)

combinations to merge two tasks out of three as shown in Fig. 4.15b–d. The CTI link usage out of 12 CTI links is different depending on the combinations. The number of unused CTI links directly affects the routability of the other charge transfer tasks.

Most importantly, merging is not free. A merged task suffers efficiency degradation due to single CTI voltage constraint. Therefore, we have to consider not only the routability but also the efficiency at same time.

4.3.4 Networked Charge Transfer Interconnect Routing

We present the proposed networked CTI routing algorithm in Algorithm 2 [8]. The input of Algorithm 2 is the CTI network G and a set of charge transfer tasks τ. Algorithm 2 iteratively performs rip-up and rerouting the charge transfer tasks until all the tasks are routed. The kernel of the routing algorithm is based on the negotiated congestion (NC) routing algorithm in [12]. The cost of resources (CTI links) gradually increases over iterations, and each charge transfer task competes with others to occupy the resource. Only one charge transfer task that is willing to pay the cost occupies the resource, and the other tasks detour via other less-costly resources.

We first define the cost of resources taking into account the distinctive characteristics of the CTI routing problem. An edge $e = (u, v)$ is associated with a congestion cost $c[e]$ that is defined as

$$c[e] = (b[e] + h[e]) \cdot p[e], \tag{4.30}$$

where $b[e]$ is the base cost of the edge e, $h[e]$ is the congestion history cost and $p[e]$ is the penalty due to the congestion at the current iteration. The base cost $b[e]$

Algorithm 2: Networked CTI routing algorithm

```
Input: CTI graph G, Charge transfer task set τ
Output: Routing tree for each task with the optimal voltage
Initialize cost c
Initialize conflict graph G^c
while shared resource exists do
    while routing retry conditions hold do
        NC-route τ on G_cti with cost c
        Update conflict graph G^c
    Solve the charge transfer optimization problem for each task
    if routing failed then
        if previous merging is not successful then
            Reject the previous merging and restore τ, G^c, and costs of E
            Mark rejected pair of tasks in G^c
        Save the current τ, G^c, and costs of E
        Merge the two tasks and update τ
        Update conflict graph G^c
        Update costs of E
```

is related with the unit cost of charge transfer from u to v, and we set the base cost to 1. The penalty $p[e]$ is defined as

$$p[e] = 1 + p_{gradient} \cdot u[e], \tag{4.31}$$

where $p_{gradient}$ is a constant, and $u[e]$ is the number of charge transfer tasks that share the edge e. The congestion history cost $h[e]$ increases gradually after each iteration to increase cost of congested edge and make the conflicting nets to avoid it. That is,

$$h[e] = \begin{cases} h[e]' & \text{if } u[e] = 0 \\ h[e]' + h_{gradient} \cdot (u[e] - 1) & \text{if } u[e] \geq 1 \end{cases}, \tag{4.32}$$

where $h[e]'$ is $h[e]$ of the previous iteration, $h_{gradient}$ is a constant, and $h[e]$ is initially 0.

Only the congestion history cost is dependent on the number of iterations by (4.32), and it is a non-decreasing function of the number of iterations. This is because the nets to be routed do not change over iterations in the signal routing, and so the congested resources are likely to be congested again in subsequent iterations. This is not the case for the CTI routing problem because we merge conflicting tasks into one, and then the shared resources are not congested any more. The cost of the previously shared edges are overestimated if we do not decrease $h[e]$ after they are merged. This leads to other charge transfers to avoid using the released edges and results in non-optimal routing results. Therefore, we reduce $h[e]$ of edges that have been congested by the merged tasks.

We define a conflict graph as $G^c = (V^c, E^c)$. There are $k = |\tau|$ nodes in $V^c = \{v_1^c, v_2^c, \ldots, v_k^c\}$, and each v_i^c is mapped to T_i. A conflict graph G^c is a complete graph, and each edge $e^c = (v_i^c, v_j^c) \in E^c$ is assigned with $d[e^c]$ which is *conflict count* between tasks T_i and T_j. Initially, $d[e^c]$ is set to zero, and we increase $d[e^c]$ by n if the tasks T_i and T_j share n CTI links. We define the sum of the conflict counts of all the edges in a conflict graph G^c to be a *conflict degree* $D[G^c]$ such that

$$D[G^c] = \sum_{e^c \in E^c} d[e^c], \tag{4.33}$$

which is the metric of routability of a given task set.

We also use this conflict graph to prune away the task pairs that do not increase the routability after merging. This is important to efficiently find task pairs to merge by avoiding a situation of trying all the pairs in every iteration. We try merging a pair of tasks, and accept it if it increases the routability. We define that the routability is improved if the conflict degree is reduced after merging by the conflict count between merged transfer tasks or more. That is, we accept the merging of T_i and T_j if

$$D[G^{c\prime}] \le D[G^c] - d[(v_i^c, v_j^c)], \tag{4.34}$$

or reject it otherwise. We mark an edge of rejected task pair with $r[e^c] = 1$ to indicate the task pair is previously rejected, and $r[e^c] = 0$ otherwise.

We merge a pair of tasks T_i and T_j that have the least difference in the optimal CTI voltage if they conflict ($d[(v_i^c, v_j^c)] > 0$) and have not been rejected previously ($r[(v_i^c, v_j^c)] = 0$). Merging the two tasks results in a new conflict graph because two tasks T_i and T_j is removed and a new task $T_{i,j}$ is added. The new task is marked not-to-be-merged ($r[e^c] = 1$) with existing tasks if both the merged tasks were marked not-to-be-merged with the tasks. The conflict count $d[e^c]$ is reset to zero after merging.

The algorithm starts from initialization of the cost of $e \in E$ based on (4.30)–(4.32) in Line 1. Initially, $u[e] = 0$ for all e. It also initializes the conflict graph G^c in Line 2. We try routing and merging until all the CTI links are not shared by multiple charge transfer tasks in the loop through Lines 3–15. The loop in Lines 4–6 attempts to route the given task set with the NC-router. The NC-router repeats rip-up and rerouting for all the charge transfer tasks while updating the edge cost $c[e]$ in Line 5. We update the conflict graph after one trial for the rip-up and rerouting for all the charge transfer tasks in Line 6. These procedures are repeated until the current task set τ is fully routed. The algorithm is terminated and returns the routing results after the charge transfer optimization for each task in Line 7 if the routing is successful.

We perform merging through Lines 8–15 if the routing fails. We judge that the task set is not routable if routing attempt fails for a certain number of iterations or a certain amount of runtime. The previous merging is rejected in Line 10 if it fails to improve the routability. If the previous merging is rejected, we restore the previous states of τ, G^c, and edge costs of E. We mark rejected pairs of tasks at the edges ($r[e^c] = 1$) in Line 11 so that they are not explored in the future attempts for merging.

Merging tasks begins with saving the current states of τ, G^c, and edge costs of E so that we can restore them when the merging is rejected in Line 12. We utilize the conflict graph G^c to find candidate tasks to be merged. We update G^c and reset the conflict count $d[e^c]$ to zero after merging in Line 14. We also update the cost c of CTI links based on the new CTI link utilization after merging in Line 15.

4.4 Experiments

4.4.1 Experimental Setup

We demonstrate examples of the proposed networked CTI architecture and evaluate the proposed CTI routing algorithm compared with the state-of-the-art shared-bus CTI architecture in this section. The proposed CTI routing algorithm is not restricted to a specific topology, but we assume a CTI network of a regular-shape mesh-grid for the demonstration purpose. All the EES banks are supercapacitor banks, and thus the terminal voltage of each bank is linearly proportional to the state of charge and is initially different to each other. The initial terminal voltage of the EES banks is randomly determined between 15 and 200 V.

The performance metric to be evaluated is the energy efficiency of charge transfer tasks. The baseline method is the shared-bus CTI architecture. We first begin with the charge transfers tasks that are single-source-single-destination (SSSD) ($|\sigma_i| = 1$ and $|\delta_i| = 1$ for all T_i). The SSSD transfers become multiple-source-multiple-destination (MSMD) transfers after merging. We do not lose any generality by assuming SSSD transfer tasks because the proposed algorithm can handle arbitrary number of nodes in σ and δ.

We assume the followings in charge transfers in the experiments. (i) An SSSD transfer task defines the amount of energy into the destination node. The amount of charge transfer is defined from the destination side in an SSSD transfer task. The amount of energy from the source node is determined accordingly by the power converter efficiency. (ii) The amount of energy into the destination nodes is kept the same in an MSMD transfer task after merging. We keep the ratio of the amount of energy to be discharged from each source the same.

4.4.2 Experimental Results

We use 3-by-3 to 7-by-7 mesh-grid CTI networks with different number of initial charge transfer tasks as benchmarks. Table 4.3 shows the number of nodes that

Table 4.3 Routing results and efficiency of charge transfers in networked CTIs. All the tasks are initially SSSD transfers

No.	Network grid size	Number of participating nodes	Number of tasks change	Energy efficiency: Networked CTI archi. (%)	Energy efficiency: Shared-bus CTI archi. (%)
1	3-by-3	4 out of 9	$2 \rightarrow 2$	88.5	76.1
2	3-by-3	8 out of 9	$4 \rightarrow 2$	79.2	73.4
3	5-by-5	12 out of 25	$6 \rightarrow 6$	81.2	74.2
4	5-by-5	18 out of 25	$9 \rightarrow 6$	57.7	57.3
5	7-by-7	18 out of 49	$9 \rightarrow 8$	81.6	74.8
6	7-by-7	38 out of 49	$19 \rightarrow 15$	75.9	68.7

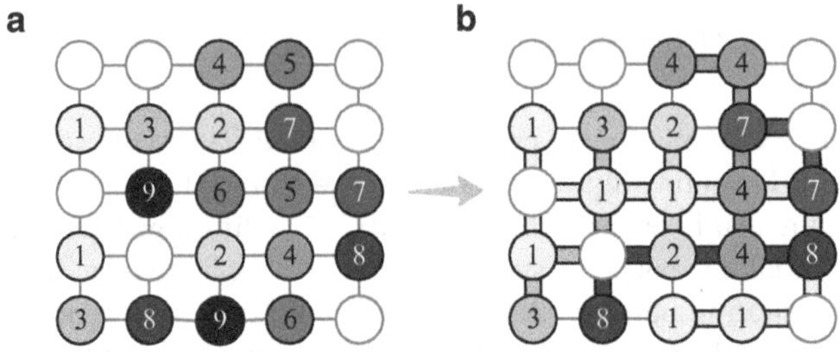

Fig. 4.16 An example of routing result of nine tasks on a 5-by-5 regular-shaped mesh-grid CTI network. (**a**) Input CTI network and task set. (**b**) Output routing tree and task set

participate in the charge transfer, total number of nodes, number of tasks in the initial and final output task sets, and the energy efficiency improvement compared with the shared-bus CTI architecture.

Figure 4.16 shows the input and output of the proposed algorithm with the benchmark No. 4 having a 5-by-5 CTI network and an initial task set of nine SSSD tasks. The CTI algorithm performs five times of routing and four times of merging (the last routing is not followed by merging). Three merges are accepted and one is rejected, and so the initial nine tasks are merged into six tasks as a result of the routing. Figure 4.16b shows that tasks T_1, T_6, and T_9 in Fig. 4.16a are merged into T_1, and tasks T_4 and T_5 are merged into T_4.

The experimental results show that the proposed routing algorithm successfully routes the charge transfer tasks even the number of tasks is large and the routing resources are limited. For example, there are initially 19 SSSD tasks in the benchmark No. 6, and so 38 nodes out of total 49 nodes participate in the charge transfers, which results in a very congested CTI network routing. The proposed algorithm merges only 4 of 19 tasks into other tasks and achieves a 7.2 % higher energy efficiency compared with the charge transfers on a shared-bus CTI architecture. The energy efficiency improvement is up to 12.4 % for the example benchmark set.

It is shown that the efficiency improvement diminishes as more number of tasks are merged. Benchmarks No. 1 and 2 end up with two tasks in the end, but efficiency improvement is less significant in No. 2 because we merge more number of tasks. It is the same for the benchmarks No. 3 and 4 that both end up with six tasks after merging. This is because more participating nodes in the same number of tasks imply that there are more nodes that do not have the optimal CTI voltage.

The energy efficiency improvement is significant when we consider the initial voltages of the EES banks are totally randomly generated. In fact, the energy efficiency improvement may be minor as in the benchmark No. 4 if the initial SSSD transfer tasks have a large voltage difference between the source and destination nodes. However, the benefit of the networked CTI architecture is larger when the voltage difference between the source and destination nodes is small in the initial charge transfer tasks before merging.

References

1. Balasubramonian R, Albonesi D, Buyuktosunoglu A, Dwarkadas S (2000) Memory hierarchy reconfiguration for energy and performance in general-purpose processor architectures. In: Proceedings of the ACM/IEEE International Symposium on Microarchitecture (Micro), pp 245–257
2. Chang YW, Thakur S, Zhu K, Wong DF (1994) A new global routing algorithm for FPGAs. In: Proceedings of the International Conference on Computer-Aided Design (ICCAD), pp 356–361
3. Chen H, Cong TN, Yang W, Tan C, Li Y, Ding Y (2009) Progress in electrical energy storage system: A critical review. Progress in Natural Science 19(3):291–312
4. Choi Y, Chang N, Kim T (2007) DC–DC converter-aware power management for low-power embedded systems. IEEE Transactions on Computer-Aided Design of Integrated Circuits and Systems 26(8):1367–1381
5. Fang X, Kutkut N, Shen J, Batarseh I (2010) Analysis of generalized parallel-series ultracapacitor shift circuits for energy storage systems. Renewable Energy
6. Hill D (1991) A CAD system for the design of field programmable gate arrays. In: Proceedings of the ACM/IEEE Design Automation Conference (DAC), pp 187–192
7. Kim Y, Park S, Wang Y, Xie Q, Chang N, Poncino M, Pedram M (2011) Balanced reconfiguration of storage banks in a hybrid electrical energy storage system. In: Proceedings of the International Conference on Computer-Aided Design (ICCAD), pp 624–631
8. Kim Y, Park S, Chang N, Xie Q, Wang Y, Pedram M (2012) Networked architecture for hybrid electrical energy storage systems. In: Proceedings of the ACM/IEEE Design Automation Conference (DAC), pp 522–528
9. Kursun V, Narendra S, De V, Friedman E (2003) Monolithic DC-DC converter analysis and MOSFET gate voltage optimization. In: Proceedings of the IEEE International Symposium on Quality Electronic Design (ISQED), pp 279–284
10. Linear Technology (2009) LTM4609: 36 V_{in}, 34 V_{out} high efficiency buck-boost DC/DC μModule regulator
11. Luu J, Kuon I, Jamieson P, Campbell T, Ye A, Fang WM, Rose J (2009) VPR 5.0: FPGA cad and architecture exploration tools with single-driver routing, heterogeneity and process scaling. In: Proceedings in the ACM/SIGDA International Symposium on Field-Programmable Gate Arrays (FPGA), pp 133–142
12. McMurchie L, Ebeling C (1995) PathFinder: a negotiation-based performance-driven router for FPGAs. In: Proceedings in the ACM/SIGDA International Symposium on Field-Programmable Gate Arrays (FPGA), pp 111–117

13. Mirhoseini A, Koushanfar F (2011) HypoEnergy. hybrid supercapacitor-battery power-supply optimization for energy efficiency. In: Proceedings of the Design, Automation & Test in Europe Conference & Exhibition (DATE), pp 1–4
14. Pedram M, Chang N, Kim Y, Wang Y (2010) Hybrid electrical energy storage systems. In: Proceedings of the ACM/IEEE International Symposium on Low-Power Electronics and Design (ISLPED), pp 363–368
15. Shin D, Kim Y, Seo J, Chang N, Wang Y, Pedram M (2011) Battery-supercapacitor hybrid system for high-rate pulsed load applications. In: Proceedings of the Design, Automation & Test in Europe Conference & Exhibition (DATE), pp 1–4
16. Stratakos AJ (1999) High-efficiency low-voltage DC-DC conversion for portable applications. PhD thesis, University of California, Berkeley, CA
17. Swartz JS, Betz V, Rose J (1998) A fast routability-driven router for FPGAs. In: Proceedings in the ACM/SIGDA International Symposium on Field-Programmable Gate Arrays (FPGA), pp 140–149
18. Wang Y, Kim Y, Xie Q, Chang N, Pedram M (2011) Charge migration efficiency optimization in hybrid electrical energy storage (hees) systems. In: Proceedings of the ACM/IEEE International Symposium on Low Power Electronics and Design (ISLPED), pp 103–108
19. Xie Q, Wang Y, Kim Y, Chang N, Pedram M (2011) Charge allocation for hybrid electrical energy storage systems. In: Proceedings of the IEEE/ACM/IFIP International Conference on Hardware/Software Codesign and System Synthesis (CODES+ISSS), pp 277–284
20. Xie Q, Wang Y, Kim Y, Shin D, Chang N, Pedram M (2012) Charge replacement in hybrid electrical energy storage systems. In: Proceedings of the Asia and South Pacific Design Automation Conference (ASP-DAC), pp 627–632
21. Zheng H, Lin J, Zhang Z, Gorbatov E, David H, Zhu Z (2008) Mini-rank: Adaptive dram architecture for improving memory power efficiency. In: Proceedings of the ACM/IEEE International Symposium on Microarchitecture (Micro), pp 210–221

Chapter 5
Joint Optimization with Power Sources

In this chapter, we discuss joint optimization of power sources and HEES systems. We first introduce a technique to maximize the energy transferred from a PV module to a HEES system by maximizing power conversion efficiency. Non-ideal characteristics of the PV modules and power converter are considered, and we show the actual stored energy is maximized by the proposed design and operation techniques. Next, we introduce a PV module emulation technique that helps the realization of the above-mentioned joint optimization of a PV module and a HEES system. Incorporating a PV module with a HEES system requires extensive experiments for determining the configuration of PV module and tuning the HEES system for maximum power generation and delivery. By accurately modeling the power generation of a PV module, the proposed PV module emulator enables rapid and efficient system development. The content of this chapter is in part based on [4, 5, 7].

5.1 Maximum Power Transfer Tracking

5.1.1 Modeling a PV Module

A PV cell converts light energy into electricity by the photovoltaic effect. This section briefly discusses the characteristics of the PV cell. It is the practice to compose PV modules with multiple, identical PV cells connected in series and/or parallel to obtain high voltage and current level. The I-V characteristic of a PV module is heavily dependent on the solar irradiance level, and the *maximum power point (MPP)* also changes significantly. A typical equivalent circuit model of a PV

Y. Kim, N. Chang, *Design and Management of Energy-Efficient Hybrid Electrical Energy Storage Systems*, DOI 10.1007/978-3-319-07281-4_5

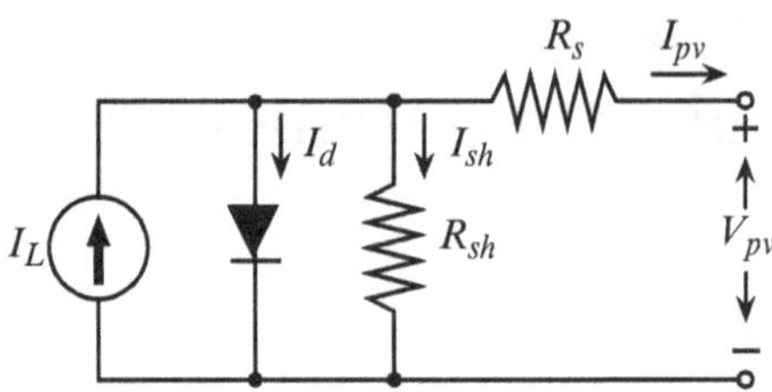

Fig. 5.1 Equivalent circuit model of a PV module

module [13] is shown in Fig. 5.1, with I-V characteristics given by:

$$\begin{aligned} I_{pv} &= I_L - I_d - I_{sh} \\ &= I_L(G) - I_0(T)\left(e^{(V_{pv}+I_{pv}\cdot R_s)(q/AnkT)} - 1\right) - \frac{V_{pv} + I_{pv}\cdot R_s}{R_p}, \end{aligned} \tag{5.1}$$

where

$$I_L(G) = \frac{G}{G_{STC}} \cdot m \cdot I_{L,cell}(G_{STC}), \tag{5.2}$$

and

$$I_0(T) = m \cdot I_{0,cell}(T_{STC}) \left(\frac{T}{T_{STC}}\right)^3 e^{(qE_g/Ank)(1/T_{STC}-1/T)}. \tag{5.3}$$

Here V_{pv} and I_{pv} are the voltage and current of the PV module, respectively. For the parameters, G is the irradiance level; T is the cell temperature; n and m are the number of connected cells in series and parallel in the PV module, respectively; q is the charge of the electron; E_g is the energy bandgap and k is Boltzmann's constant. STC stands for standard test condition in which irradiance level is 1,000 W/m^2 and temperature is 25 °C. The following five parameters determines the characteristics of the PV module.

- $I_{L,cell}(G_{STC})$: photo-generated current of a cell at standard test condition.
- $I_{0,cell}(T_{STC})$: dark saturation current of a cell at standard test condition.
- R_s: equivalent module series resistance.
- R_{sh}: equivalent module parallel (shunt) resistance.
- A: the diode ideality factor.

We use a method proposed in [12] to extract the above-mentioned five parameters from datasheet values in STC, which consist of the open circuit voltage V_{oc}, the short circuit current I_{sc}, voltage V_{mpp} and current I_{mpp} at the MPP, and temperature coefficients. Then a set of five parameters that determine V_{pv}–I_{pv} characteristic are derived from the measured G and T by (5.1)–(5.3).

Figure 5.2 shows the variation of I_{pv} and P_{pv} by V_{pv} of a PV module with around 30 W power capacity at G_{STC} = 1,000 W/m^2. It shows I_{pv}–V_{pv} curves and

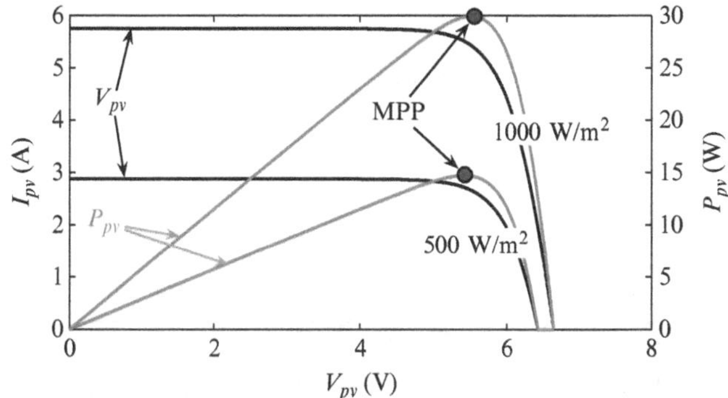

Fig. 5.2 An example of I_{pv} and P_{pv} variation by V_{pv} when $G = 500\,\text{W/m}^2$ and $G = 1{,}000\,\text{W/m}^2$

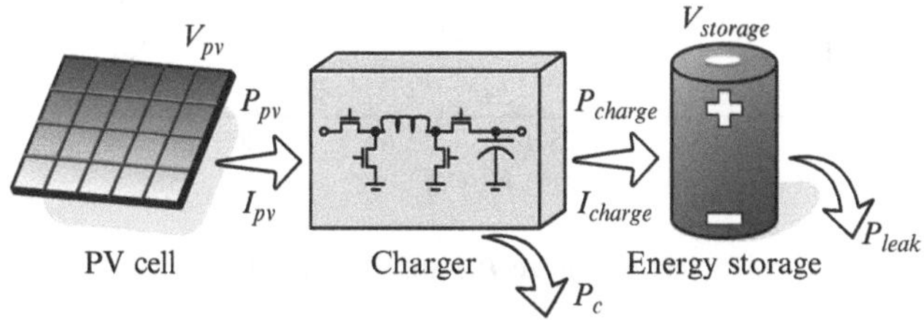

Fig. 5.3 Power transfer and loss in a solar energy harvesting system

P_{pv}–V_{pv} curves at two different G of 500 and 1,000 W/m^2. We see that the curves significantly change depending on the solar irradiance levels. The *maximum power point tracking (MPPT)* methods adjust the operating point, which is defined as a pair of (V_{pv}, I_{pv}), to (V_{mpp}, I_{mpp}) against dynamically varying irradiance. Also, we see that both V_{pv} and P_{pv} change in a wide range depending on I_{pv} even with the same irradiance level.

5.1.2 Maximum Power Transfer Point

5.1.2.1 Sub-optimality of Maximum Power Point Tracking

Figure 5.3 shows the architecture of the solar energy harvesting system. The conventional MPPT techniques aim at maximizing P_{pv}. However, a solar energy harvesting system has a charger circuit which involves a non-zero amount of energy loss P_c, and so P_{charge} is not equal to P_{pv} as we have seen from (5.4). Furthermore, P_c is not constant as we discussed in Sect. 4.1. This results in maximizing P_{pv} does not necessarily guarantee the maximum P_{charge}, and thus the MPPT does not always achieve the maximum harvested energy.

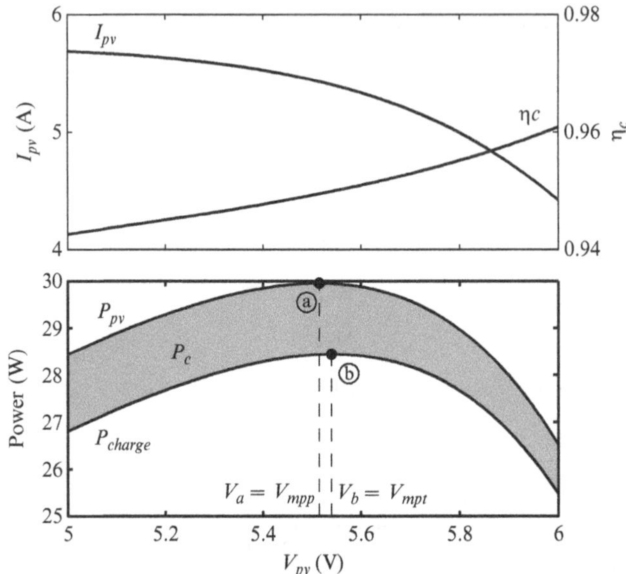

Fig. 5.4 Different V_{pv} points that maximize P_{pv} and P_{charge} when efficiency varies ($P_{charge} = \eta_c \cdot P_{pv}$)

Figure 5.4 shows an example of such a phenomenon. It shows variations of P_{pv}, P_{charge}, and η_c, whose relationship is shown in (5.4), depending on V_{pv}. Conventional MPPT simply finds the MPP (V_{mpp}, I_{mpp}) that results in $P_{pv} = P_{mpp}$ without consideration of the charger efficiency η_c at that point. In Fig. 5.4, P_{pv} is maximized at ⓐ when $V_{pv} = V_a$. However, η_c is not high at the MPP, and this results in lower P_{charge} than when $V_{pv} = V_a$ where η_c is higher. Rather, P_{charge} at ⓑ is higher than P_{charge} at ⓐ thanks to higher η_c even though P_{pv} at this point is smaller than P_{mpp}. This is more desirable to operate at ⓑ in that we are interested in maximizing P_{charge} rather than P_{pv} in practice. Consequently, the MPPT and maximum efficiency tracking of the charger should be considered at the same time to overcome the sub-optimality of the MPPT. We have to take the dynamic status of the energy storage device into account when finding the optimal operating point that maximizes P_{charge} because η_c is affected by them. Therefore, we design the system to maximize the output of the charger P_{charge}, not the input of the charger P_{pv}, in order to achieve the system-level energy optimum.

5.1.2.2 Maximum Power Transfer Tracking

We first define the *maximum power transfer (MPT)* point, in contrast to the MPP, as the operating point where the P_{charge} is the maximum [4]. We name the voltage and current of the PV module at the MPT point V_{mpt} and I_{mpt}, respectively. We define P_{mpt} as the maximum P_{charge} that is available at the MPT point (V_{mpt}, I_{mpt}). That is,

$$P_{mpt} = V_{mpt} \cdot I_{mpt} - P_c. \tag{5.4}$$

For a given PV module, P_{mpt} is a function of G and $V_{storage}$ when the charger efficiency characteristics are given because P_{charge} is dependent on η_c, whereas P_{mpp} is solely dependent on G. The MPT point changes over time depending on the environmental conditions such as G and system condition such as $V_{storage}$ as mentioned above. The *maximum power transfer tracking (MPTT)* is a run-time operational method that dynamically adjusts the operating point to (V_{mpt}, I_{mpt}). We accomplish the MPTT by measuring P_{charge} and controlling the charger to maximize it. We may use the conventional tracking methods of the MPPT such as the perturb-and-observe or incremental conduction for the MPTT as well.

Figure 5.2a shows V_{pv} and P_{pv} variations according to I_{pv} with four different irradiance values. Even with the same irradiance, we can see a significant change in P_{pv}. The MPPT methods discussed in Sect. 5.1.2.1 can be used to find the maximum P_{pv} regardless of the irradiance.

However, if we take the charger efficiency into account, the P-I curve we have to consider is the P_{charge}-I_{charge} curve shown in Fig. 5.2b, rather than the P_{pv}-I_{pv} curve of Fig. 5.2a. Since the x-axis is I_{charge}, the y-axis P_{charge} is linearly proportional to the I_{charge} when V_{cap} is given. The maximum P_{charge} values, marked by squares, are the MPT points. Beyond this point, further increment of I_{charge} causes a rapid drop of V_{pv}. The V_{pv} values at the MPT points are not the same as the V_{pv} values at the MPP, and P_{charge} at the MPT point is slightly lower than P_{pv} at the MPP because $\eta < 1$.

Our proposed MPTT keeps tracking (V_{pv}, I_{pv}) which may be slightly different from that of conventional MPPT, to guarantee the maximum amount of power transferred to the load at all times rather than the maximum power extracted from the PV module. Note that the proposed MPTT always outperforms the conventional MPPT in terms of net energy delivery to the load regardless of environmental conditions.

P_{charge} at the MPT is determined not only by G, but also by the current value of V_{cap}. Figure 5.5a is the surface that consists of the maximum P_{charge} values of a 7×10 PV array in the $G - V_{cap}$ domain. We may draw a trace of $\left(V_{cap}(t), G(t)\right)$ pairs on this surface with the solar irradiance set to a meaningful value in $t \in [t_{sunrise}, t_{sunset}]$. For instance, the white lines are the traces when C is 300, 3,000 or 30,000 F. For illustration purpose, we set $G(t_{noon}) = 900\ \mathrm{W/m^2}$. Initially, $V_{cap}(t_{sunrise}) = 0$ V. From (5.9), the value of $P_{charge} - P_{leak}$ is the gradient of E_{cap}. Figure 5.5b shows the supercapacitor's energy, E_{cap} as t elapses. Right after the sun rises, G is low and P_{charge} is also low, and E_{cap} increases slowly. Generally, with a reasonable supercapacitor of C, P_{charge} has the maximum values during the day, and E_{charge} increases most rapidly at noon ($t = t_{noon}$). The sampling points in Fig. 5.5a, b are matched with each other in terms of t. E_{cap} slightly decreases in the evening due to leakage of the supercapacitor such that $P_{leak} > P_{charge}$ in the evening.

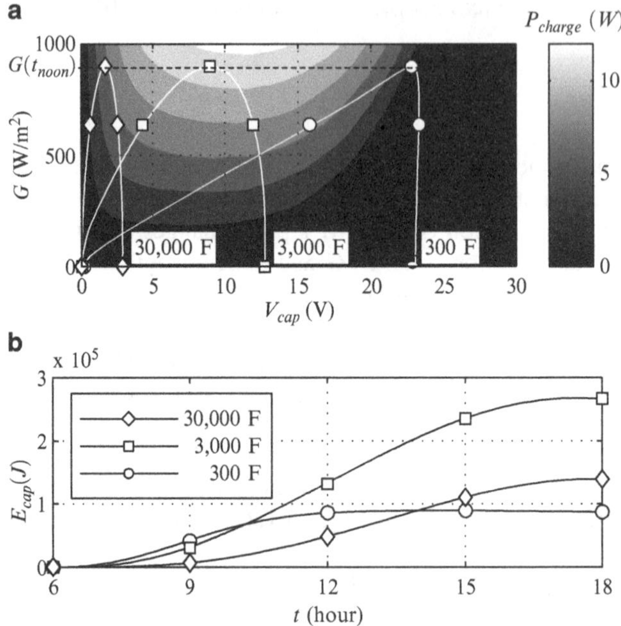

Fig. 5.5 Energy accumulation trace of a 7 × 10 PV array. (**a**) P_{charge} and $\left(V_{cap}(t), G(t)\right)$ trace line. (**b**) E_{cap} change over t

5.1.3 MPTT-Aware Energy Harvesting System Design

5.1.3.1 Optimal System Design Problem

We harvest the solar energy during daytime and typically store some of the energy for use during the nighttime. To focus on energy storage efficiency and avoid divergence from the main context, we assume that there is no energy usage during the daytime. The amount of energy to be stored at the end of the day, E_{req} is given as a design requirement. The objective of the proposed design optimization is to derive a cost-effective and energy-efficient design of a solar energy generation and storage system to achieve this requirement.

Cost-effectiveness is defined as the minimum number of PV modules that meet the energy storage requirement, E_{req}. Energy efficiency means maximizing the amount of energy that is eventually stored in the supercapacitor at the end of the day with the same number of PV modules. These two optimizations are coupled together; if we enhance the efficiency, we may use small number of PV modules while satisfying E_{req}.

The amount of energy stored during the daytime is given by

$$E_{cap}(t_{sunset}) = E_{cap}(t_{sunrise}) + \int_{t_{sunrise}}^{t_{sunset}} \left(P_{charge}(t) - P_{leak}(t)\right) dt. \tag{5.5}$$

We must consider another constraint, i.e., feasibility. Commercially available switching converters and chargers have the maximum voltage rating around 30 VDC, and we must thus limit the maximum voltage of the supercapacitor bank and the PV modules in series. We consider identical PV arrays with the configuration of n in series and m in parallel. Due to this regularity, we can operate all the PV cells at the identical operating condition and maintain uniform energy efficiency for all cells.

The maximum amount of energy that can be stored in a supercapacitor bank is proportional to $C \cdot {V_{cap\_max}}^2$, where V_{cap_max} is the maximum voltage rating of the supercapacitor bank. We can change C and V_{cap_max} by connecting more unit-sized supercapacitors in parallel and series, respectively. However, the total number of unit-sized supercapacitors necessary to store a certain amount of energy does not change no matter how we configure the series and parallel connections. Therefore the supercapacitor cost, which is proportional to the total number of unit-sized supercapacitors, is not the optimization objective that we consider. We first derive the amount C that results in maximum energy harvesting for the given E_{req}, and next, we determine V_{cap_max} from

$$V_{cap}(t) = \frac{Q(t)}{C} = \sqrt{\frac{2 \cdot E_{cap}(t)}{C}}. \tag{5.6}$$

To make a long story short, smaller voltage difference between the PV array and supercapacitor bank achieves better charger efficiency. Thus the key optimization method is to control the supercapacitor voltage by adjusting C, because the PV array voltage is determined by solar irradiance, which is not controllable. For example, if the supercapacitor bank capacitance is too small, the supercapacitor bank voltage rises quickly and goes way higher than the nominal PV array voltage. On the other hand if the supercapacitor bank capacitance is too big, the supercapacitor bank voltage does not rise much and remains far lower than the nominal PV array voltage. Determination of the supercapacitor bank capacitance is also coupled with the PV array configuration. We must jointly optimize how many series and parallel connections of PV cells within a PV array should be established given the supercapacitor capacitance.

We name the overall PV-supercapacitor system efficiency optimization as MPTT design. The MPTT design can be formally described as follows:

- Given: The energy requirement E_{req} and the maximum voltage rating V_{rating}.
- Prerequisite: PV module characteristics, charger efficiency model, and solar irradiance profile G.
- Objective: (i) Find a $n \times m$ PV array configuration which minimizes $n \cdot m$ while meeting $E(t_{sunset}) \geq E_{req}$; and (ii) given the $n \times m$ configuration, find a supercapacitor bank capacitance C that maximizes the energy efficiency while meeting $V_{cap}(t_{sunset}) \leq V_{rating}$.

It is possible to have $V_{cap}(t)$, for some t, higher than $V_{cap}(t_{sunset})$ due to the self-discharge. However, we assume this is negligible, as we will see in the following section for a half-day storage.

5.1.3.2 Design Optimization

PV Array and the P_{charge} Surface

The configuration of the PV array determines the shape and magnitude of the P_{charge} surface in the $G - V_{cap}$ domain. The total number of PV modules is $N = n \cdot m$, which determines the magnitude of the P_{charge} surface. Evidently the more PV modules are used, the higher power can be achieved. We define P_{peak} and V_{opt} of a P_{charge} surface as

$$P_{peak} = \max_{\forall V_{cap}} \left(P_{charge} \left(G(t_{noon}), V_{cap} \right) \right), \quad (5.7)$$

$$V_{opt} = \arg\max_{V_{cap}} \left(P_{charge} \left(G(t_{noon}), V_{cap} \right) \right), \quad (5.8)$$

which gives the maximum possible power that goes to the supercapacitor and its corresponding condition, with an $n \times m$ PV array. The values of n and m determine the location of P_{peak} and V_{opt}. As n increases, V_{opt} increases almost linearly. To achieve the maximum P_{charge} when G is the maximum ($t = t_{noon}$), $V_{cap}(t_{noon})$ should be equal to V_{opt} such that $P_{charge}(t_{noon}) = P_{peak}$.

Supercapacitor Size and Harvested Energy

It is shown in Fig. 5.6 that V_{cap} is a critical factor for P_{charge}. From (5.6), V_{cap} is inversely proportional to C. Therefore the determination of C is very important for maximizing P_{charge}, and in turn, the accumulated energy, E_{cap}. Figure 5.7 shows the total amount of accumulated energy, $E_{cap}(t_{sunset})$, as a function of C. It turns out that neither a small nor a large C is energy efficient. This implies that an ad-hoc decision on C may result in a poor energy harvesting. Each curve in Fig. 5.5a gives us more intuition about this result. A 3,000 F C is close to the energy optimal. If C is too small (300 F), V_{cap} increases too rapidly and V_{cap} is very different from V_{opt} when G is the maximum. Thus, the curve cannot arrive in the high P_{charge} region. On the other hand, if C is too large (30,000 F), V_{cap} increases too slowly and V_{cap} is again very different from V_{opt} when G is the maximum. Thus again the curve cannot arrive in the high P_{charge} region.

Therefore it is important to determine C so that the system remains in the high-P_{charge} region for a longer period of time. For a symmetrical irradiation profile in a day, we make $P_{charge}(t_{noon}) = P_{peak}$ by adjusting C to guarantee $V_{cap}(t_{noon}) = V_{opt}$. Based on the fact that $V_{cap}(t_{noon}) = V_{opt}$ and using the estimated $E_{cap}(t_{noon})$, the energy-optimal C is calculated by (5.6).

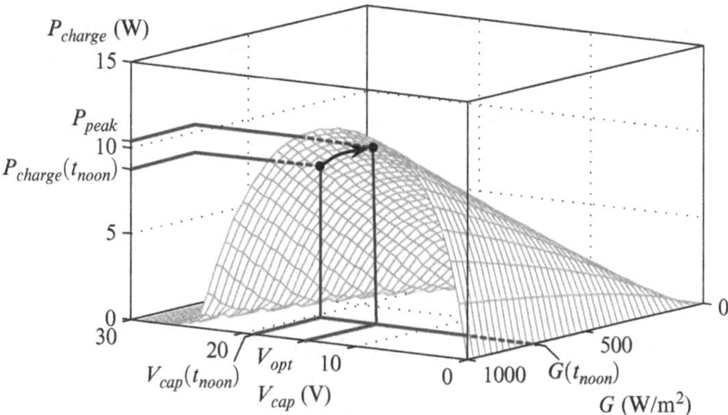

Fig. 5.6 Maximum P_{charge} surface of a 7×10 PV array in GV_{cap} domain

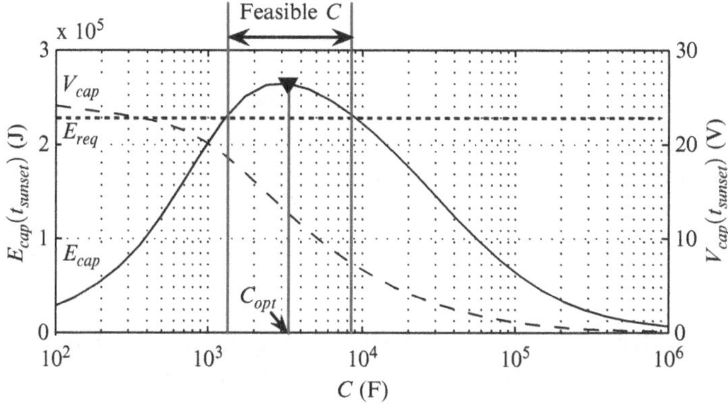

Fig. 5.7 Accumulated energy $E_{cap}(t_{sunset})$ and corresponding $V_{cap}(t_{sunset})$ of a 7×10 PV array

5.1.3.3 Systematic Design Optimization

A naive brute force method to find the optimal design is that first we obtain P_{charge} surfaces for all $n \times m$ PV array configurations, and then evaluate E_{cap} for all C. However, obtaining P_{charge} surfaces as in Fig. 5.6 is very time consuming because of the numerical iterations needed to reach the convergence point of the PV model and charger models. Furthermore, as can be seen in Fig. 5.5a, each curve passes only a part of the surface, which makes it pointless to calculate P_{charge} for all (V_{cap}, G) pairs.

Based on the observation that a switching converter exhibits a higher efficiency when the input and output voltages are similar to each other, we develop Algorithm 3 that efficiently derives the near-optimal values of n, m, and C when E_{req}, V_{rating}, and G are given. The objective of this algorithm is to derive the minimum $n \times m$ and optimal C. Since the supercapacitor cost is determined by E_{req}, C is not to

Algorithm 3: MPTT-aware solar energy harvesting system design

Input: (energy requirement E_{req}, voltage rating V_{rating}, irradiance profile G, theoretical maximum energy E_{mpp}, noon time t_{noon})

Output: (optimal PV module size n_{opt} and m_{opt}, optimal capacitance C_{opt}), or `null` if given invalid parameters

1 $N \leftarrow \lceil E_{req}/E_{mpp}\,(G(t_{noon}))\rceil$; $n_{max} \leftarrow \infty$; $m_{max} \leftarrow \infty$
2 **repeat**
3 $\quad V_{opt}(1,N) \leftarrow$ `find_V_opt`$(1,N)$; $V_{opt}(N,1) \leftarrow$ `find_V_opt`$(N,1)$ $N_{next} \leftarrow \infty$
$\quad$ **foreach** $(n,m) \in S = \{(n,m)|(n < n_{max} \vee m < m_{max}) \wedge n\cdot m = N \wedge n,m \in \mathbb{N}\}$ **do**
4 $\quad\quad V_{opt}(n,m) \leftarrow$ `linear_approx`(V_{opt}, n, m)
5 $\quad\quad P_{peak}(n,m) \leftarrow$ `maximum_P`$(V_{opt}(n,m), G)$
6 $\quad\quad E_{cap}(n,m) \leftarrow$ `estimate_E_cap`$(P_{peak}(n,m), G)$
7 $\quad\quad C(n,m) \leftarrow 2\cdot\left(E_{cap}(n,m)/2\right)/V_{opt}(n,m)^2$
8 $\quad\quad V_{cap}(n,m) \leftarrow \sqrt{2\cdot E_{cap}(n,m)/C(n,m)}$
9 $\quad\quad$ **if** $V_{cap}(n,m) > V_{rating}$ **then**
10 $\quad\quad\quad n_{max} \leftarrow \max(n_{max}, n)$; $m_{max} \leftarrow \max(m_{max}, m)$
11 $\quad\quad\quad$ **if** $m = N$ **then**
12 $\quad\quad\quad\quad$ **return** `null`
13 $\quad\quad$ **else if** $E_{cap}(n,m) < E_{req}$ **then**
14 $\quad\quad\quad N_{next} \leftarrow \min\left(N_{next}, N + \lceil N\cdot(1 - E_{cap}(n,m)/E_{req})\rceil\right)$
15 $\quad N \leftarrow N_{next}$
16 **until** $\exists(n,m)$ *such that* $E_{cap}(n,m) \geq E_{req}$ *and* $V_{cap}(n,m) \leq V_{rating}$
17 $(n_{opt}, m_{opt}) \leftarrow$ `maximum_E_cap`(E_{cap}) ; $C_{opt} = C(n_{opt}, m_{opt})$
18 **return** $(n_{opt}, m_{opt}, C_{opt})$

be minimized, but to be optimized for harvesting the largest amount of energy. This algorithm requires that the PV model and the charger efficiency model be characterized *a priori*.

We first calculate the minimum feasible number of PV modules by dividing E_{req} by E_{mpp}, which is the theoretical maximum energy that can be extracted by MPP with the maximum G (Line 1). And temporary variables are initialized subsequently. We find V_{opt} values for the two extreme cases such that all PV modules are connected in parallel or in series (Line 3).

Based on the observation in Sect. 5.1.3.2, we may linearly approximate V_{opt} for other $n \times m$ configurations from the two extreme cases (Line 4). We estimate the total harvested energy using

$$E_{cap}(t + \Delta t) = E_{cap}(t) + \Delta t \cdot \left(P_{charge}(t) - P_{leak}(t)\right) \tag{5.9}$$

and $P_{charge}(t) = P_{peak} \cdot G(t)/G(t_{noon})$ where P_{peak} corresponds to the approximated V_{opt}, considering leakage (Line 6). The corresponding C and the maximum V_{cap} are derived by equations on Lines 7 and 8. In each iteration, we prune a large portion of possible configurations that do not satisfy the voltage rating constraint from the current N and larger N (Lines 9). Even the all-parallel configuration may have V_{cap} that exceeds V_{rating} if the given V_{rating} is too low. We stop iteration in such a case (Line 11). If E_{cap} is less than E_{req}, we increase N by the ratio of insufficient energy (Line 14). This is repeated until we find a feasible configuration (Line 16). We choose the configuration that has the maximum E_{cap} among all feasible configurations (Line 17). This algorithm is scalable enough to accommodate large-scale applications due to judicious calculation of N as well as effective pruning.

5.1.3.4 Energy Harvesting Improvement

We use Linear Technology LTC3531 buck-boost converter as the charger model, and the Spectrolab GaAs/Ge single junction PV module of $A = 10\,\text{cm}^2$, which has $V_{oc} = 1.025\,\text{V}$, $I_{sc} = 0.305\,\text{A}$, and $P_{mpp} = 0.257\,\text{W}$ at $G = 1{,}353\,\text{W/m}^2$ in the experiment. We assume that, without loss of generality, the sun rises at 6:00 and sets at 18:00, and $G(t_{noon}) = 900\,\text{W/m}^2$. We assume 10 % self-discharge rate per day for the supercapacitor, which is a typical value for commercially available supercapacitors.

First we show the energy efficiency of the proposed MPTT method compared with conventional MPPT method. Table 5.1 shows the accumulated energy E_{cap} at the end of the day for various PV array and supercapacitor configurations, which are operated by the MPPT and MPTT methods. The capacitance C of the MPTT case is the theoretical optimum for each given $n \times m$. Most importantly, conventional MPPT methods have no concept of the efficiency-optimal C and PV array configuration, and any C value and any PV array configuration are supposed to yield the same amount of harvested energy. Thus, it is not surprising to have a C value and a PV array configuration for conventional MPPT that yield very poor charging efficiency. We compare the proposed MPTT with conventional MPPT for different C values and PV array configurations. When using the optimal capacitance, MPTT shows more than 6× harvested energy over a poorly configured conventional MPPT as Table 5.1 shows. The results for poorly configured MPPTs are not embellished because conventional MPPT does not care about the charger loss caused by improper C value. More interesting result is that even the accidentally optimal

Table 5.1 Energy efficiency of the conventional MPPT and suggested MPTT methods

$n \times m$	Tracking method	C (F)	V_{cap} (V)	E_{cap} (J)	Normalized E_{cap} (%)
5 × 5	MPTT	2,378	9.0	96,342	100.0
	MPPT	2,378	8.9	93,451	97.0
		23,780	2.2	59,170	61.4
		238	11.0	14,404	15.0
12 × 2	MPTT	874	15.2	101,545	100.0
	MPPT	874	14.8	95,795	94.3
		8,740	4.2	77,261	76.1
		87	19.1	15,823	15.6

Table 5.2 Energy harvesting result of designs by the proposed method and exhaustive search (ES)

E_{req} (J)	V_{rating} (V)	Opt. method	$n \times m$	C_{opt} (F)	E_{cap} (J)	E_{cap} error (%)
50k	10	Algorithm 3	6 × 2	1,159	49,507	−0.99
		ES	7 × 2	1,289	58,439	+16.88
50k	30	Algorithm 3	12 × 1	415	50,970	+1.94
		ES	12 × 1	524	51,711	+3.42
100k	30	Algorithm 3	12 × 2	648	100,011	+0.01
		ES	12 × 2	874	101,545	+1.55
200k	20	Algorithm 3	10 × 5	2,691	197,713	−1.14
		ES	10 × 5	1,713	201,702	+0.85

configuration of conventional MPPT is up to 5.7 % less efficient than the proposed MPTT. This is because MPTT finds the true optimal tracking point considering the charger loss while conventional MPPT draws more power from the PV array but loses even more power in the charger. The energy-efficient configuration of MPPT is coincidence and hard to achieve because conventional MPPT gives no clue for the optimal configuration. Note that the 5 × 5 configuration harvests less energy than the 12 × 2 configuration while using one more PV cell even the MPTT is applied, and therefore it is not an optimal design. Algorithm 3 can be used to effectively find the optimal design avoiding such a case.

We show the accuracy of Algorithm 3 in terms of actual cost and energy. Recall that Algorithm 3 tries to find the near-optimal value by a heuristic approach to make the computational complexity reasonable. Table 5.2 is the comparison between designs derived by the suggested algorithm and the optimal design found by exhaustive search for various E_{req} and V_{rating} values. For all cases, we notice that the negative error is less than 2 %, which is quite reasonable in light of the typical device tolerance used in commercial circuits. This error is mainly due to the fact that the estimation of E_{cap} is based on the observation that the trace on the P_{charge} curve may be approximated by a sinusoidal waveform. With this approximation, we cannot guarantee a positive or a negative bound on the error.

We confirm that the proposed MPTT design may have a slightly smaller E_{cap} than E_{req}. This is mainly because the curve on the P_{charge} surface is not an exact sine function. We can mitigate this error by a minor overdesign of E_{req}, which is anyway required due to the component tolerance in a real system.

5.2 Photovoltaic Emulation for MPTT

5.2.1 Model Parameter Extraction

We may extract five unknown parameters $I_L(G_0)$, $I_0(T_0)$, R_s, R_p and N from I_{sc}, V_{oc}, V_{mpp} and I_{mpp} for each I-V curve measured under a specific environmental condition (G_0, T_0) using a similar method as [12]. This conventional parameter extraction heuristic can be accelerated by Newton-Raphson iteration method. Although such method is quite stable and does not rely heavily on initial values for the iteration procedure, it does not utilize the whole I-V curve for parameter extraction, and therefore the overall average fitting error cannot be guaranteed to be minimized. Nevertheless, it is still necessary to collect the empirical data for the whole I-V curve in order to find out the MPP, unless we only can rely on the datasheet provided by the manufacturer. In fact, the fitting errors can be significant in some specific PV module I-V ranges. Hence, such method cannot fulfill the requirement of state-of-the-art researches [1, 4, 9] that the whole operating range should be accurately modeled for maximizing the energy efficiency. On the other hand, we adopt a nonlinear curve fitting algorithm here to overcome the shortcoming of the previous method that only some specific points of the whole I-V curve have been used. The parameter extraction is performed only one time for each PV module at the characterization step, and so the computational overhead for the curve fitting is negligible.

The fitting parameters depend heavily on the initial values. If the initial values are not properly set, the fitting results obtained may be not optimal nor even feasible. This is because of the fact that nonlinear curve fitting is a highly non-convex optimization problem, and it is likely to be stuck at a local optimal point. Therefore, we propose to use the parameter extraction heuristic that uses specific points of each I-V curve, accelerated by Newton-Raphson method, in the initial phase. The derived five parameters, i.e., $I_L(G_0)$, $I_0(T_0)$, R_s, R_p and N, serve as the proper initial values in the subsequent least-squares nonlinear curve fitting method based on Levenberg-Marquardt algorithm. Furthermore, we have to also set an upper bound and a lower bound of the fitting parameters, since such bounds also play an important role in the nonlinear curve fitting for acceleration and convergence. One simple, yet effective set of bounds is given by $[\alpha\mathbf{P}, \beta\mathbf{P}]$, where $\mathbf{P} = (I_L(G_0), I_0(T_0), R_s, R_p, N)$ is the derived PV module parameters in the initial phase. With such properly set initial values and upper/lower bounds, nonlinear curve fitting algorithm can find the optimal PV module parameters effectively, taking into account the whole I-V operating range.

We apply the proposed combined parameter extraction method on the measured PV module I-V curves. Significant reduction in RMS fitting error is observed compared with the conventional method which only considers some specific points.

First, we show the model accuracy improvement by the proposed characterization method. Table 5.3 shows the equivalent circuit model parameters introduced in Sect. 5.1.1 extracted by the conventional method and the proposed method, respectively. We apply these parameters to the PV model and derive the I-V curves shown in Fig. 5.8a. It shows the I-V curves obtained from the measurement compared with the I-V curves derived by the conventional and proposed characterization methods, for three different G values: 840, 730, and 590 W/m^2 at the same temperature $T = 27\,°\text{C}$. We see that the proposed method extracts more accurate parameters than the conventional method does. Fig. 5.8b–d provides a more detailed view of the curves, in a low-voltage range near short-circuit state, a medium-voltage range near the MPP state, and a high-voltage range near open-circuit state, respectively.

Table 5.3 Model parameters extracted by the conventional (NR) method and proposed (CF) method, and derived some V, I, and P compared with measured data (Meas.). $T_0 = 27\,°\text{C}$

	Parameters			Derived V (V), I (A), and P (W)			
G_0	Value	NR	CF	Value	Meas.	NR	CF
840	$I_L(G_0)$	4.66 mA	4.64 mA	V_{oc}	22.50	22.51	22.49
	$I_0(T_0)$	11.52 nA	11.52 μA	I_{sc}	3.85	3.85	3.87
	R_s	689 mΩ	405 mΩ	V_{mpp}	17.14	17.12	17.04
	R_p	41.13 Ω	58.82 Ω	I_{mpp}	3.22	3.23	3.24
	N	37.00	57.08	P_{mpp}	55.25	55.25	55.19
730	$I_L(G_0)$	4.68 mA	4.66 mA	V_{oc}	22.23	22.23	22.23
	$I_0(T_0)$	7.55 nA	6.98 μA	I_{sc}	3.37	3.36	3.38
	R_s	761 mΩ	418 mΩ	V_{mpp}	16.99	17.02	17.02
	R_p	49.63 Ω	63.91 Ω	I_{mpp}	2.84	2.83	2.83
	N	36.00	54.86	P_{mpp}	48.24	48.25	48.14
590	$I_L(G_0)$	4.87 mA	4.86 mA	V_{oc}	22.01	22.01	22.01
	$I_0(T_0)$	7.61 nA	2.03 μA	I_{sc}	2.83	2.83	2.84
	R_s	776 mΩ	472 mΩ	V_{mpp}	17.06	17.04	17.01
	R_p	53.35 Ω	62.68 Ω	I_{mpp}	2.37	2.37	2.37
	N	36.00	50.28	P_{mpp}	40.38	40.39	40.29

The I-V curves derived by the proposed method are very close to the measured data, but the others derived by the conventional method have noticeable discrepancy.

Figure 5.8c shows an interesting result regarding the MPP. The MPPs derived by the conventional method are almost the same as the measured data. It is not surprising because what this method does is to fix V_{oc}, I_{sc}, V_{mpp}, and I_{mpp} as measured and find the parameters accordingly. This implies that the conventional method highly weights these points than other points. However, as mentioned earlier, not only the accuracy nearby the MPP, but overall accuracy across the whole I-V range should not be excluded when we explore the system in terms of energy efficiency. We do not weight any certain point or range of the I-V curve in the curve fitting, to enhance the overall accuracy. The MPP derived by the proposed method is slightly shifted from the measured data, but accuracy enhancement is observed in the entire range.

It is more clearly seen in Fig. 5.9, which presents the relative and RMS errors of the derived I-V curves compared with the measured data. The curves show the relative error between the measured I_{pv} and modeled I_{pv} at $V_{pv} \in [0, V_{oc}]$ when $G_0 = 840\,\text{W/m}^2$ and $T = 27\,°\text{C}$. The bars indicate the RMS errors in ten uniform intervals of V_{pv}. The error rate of the proposed method is almost negligible when compared with the conventional method across the whole range of V_{pv}. It is noticeable that the error rate of the conventional method is small enough only nearby the MPP.

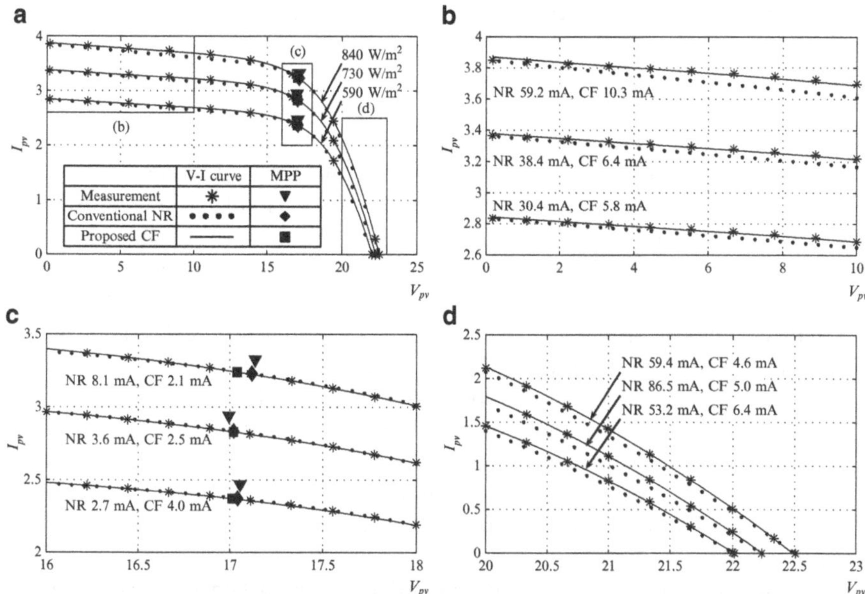

Fig. 5.8 Model comparison with I-V curves in various irradiations. Numbers associated with curves denote the RMS error in each displayed voltage range. $T = 27\,^{\circ}\mathrm{C}$. (**a**) Overall range. (**b**) Low-voltage range near short-circuit state. (**c**) Medium-voltage range near MPP state. (**d**) High-voltage range near open-circuit state

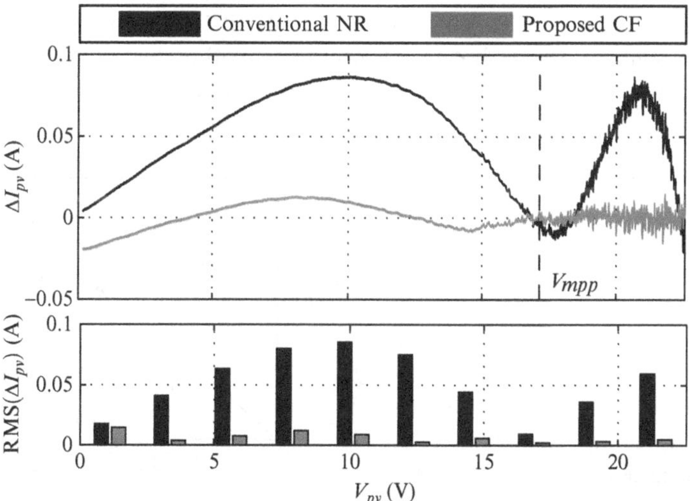

Fig. 5.9 Model errors compared with the measured data. $G = 840\,\mathrm{W/m^2}$, and $T = 27\,^{\circ}\mathrm{C}$

5.2.2 Dual-Mode Power Regulator with Power Hybridization

5.2.2.1 PV Module I-V Characteristics

Typical I-V characteristics of a PV module are shown in Fig. 5.2. A PV module basically is a current source as shown in Fig. 5.1, and forward biasing of the diode limits the output voltage which results in properties of a voltage source. An ideal PV module has a zero R_s and an infinite R_p, but a practical PV module has a non-zero R_s and a finite R_p. This non-ideal series and parallel resistances determine the gradients on the I-V curve. As mentioned above, the PV module exhibits dual behaviors which can be either a voltage source or a current source depending on the operating range. More specifically, the PV module essentially behaves as a voltage source (i.e., it supplies a constant voltage regardless of the output current) when current is low and voltage is high, and behaves as a current source (i.e., it supplies a constant current regardless of the output voltage) when voltage is low and current is high. The boundary between the voltage source region (VS)R and current source region (CSR) is not very definitive, but reference [2, 14] defines it to be the MPP of the PV module.

Consequently, the PV module shows different output behavior even with the same amount of load power variation by its operation regions. Figure 5.10 illustrates how the load power variation affects the output of the PV emulator in the two different regions. The load power variation (①) results in a small voltage variation (②) and a large current variation (③) in the VSR. In contrast, the same amount of load power variation results in a small current variation (⑤) and a large voltage variation (④) in the CSR. This implies that the PV emulators sorely based on a voltage regulator should be able to react to a small current change with a high feedback control gain in the CSR, which may result in instability in the VSR.

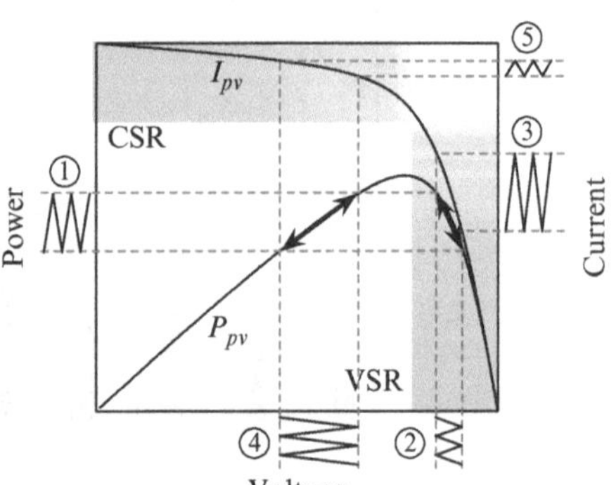

Fig. 5.10 Voltage and current variations in the VSR and CSR

5.2.2.2 Modes of Operation

In spite of the dual characteristics of the PV module's output, previous PV emulators have relied on only a voltage regulator to reproduce the complete output I-V curve of the PV module. However, in the CSR, change in the output voltage of the PV module induced by the load impedance variation can be quite large. Under these conditions, current-based control provides a better control quality, and in turn, higher PV emulation accuracy in the CSR, in the same way that voltage-based control is preferred in the VSR. Nevertheless, we do not want to use a current regulator to reproduce the entire output I-V curve of the PV module because it may result in low accuracy in the VSR. Even if a nested feedback controller is used (e.g., an outer voltage control loop and an inner current control loop, or vice versa) [6, 11], the PV emulator will exhibit poor output controllability (and hence poor emulation accuracy) in one or the other of the two regions of operation.

Therefore, in order to reproduce the original characteristics in both the VSR and CSR, we use two separate power sources [5]. In particular, we use a voltage regulator to generate a regulated voltage when the target PV module is operating in the VSR, and use a current regulator to generate regulated current in the CSR. We call the two operating modes of the dual-mode power regulator *voltage regulation mode (VRM)* and *current regulation mode (CRM)*. It is required to develop an elaborate power hybridization circuit which supports the VRM and CRM for the implementation of the PV emulator. A sophisticated control method also has to be designed in order to seamlessly switch between the two operating modes.

It is essential for a PV emulator to supply uninterrupted power to the load device, and so at least one of the voltage and current regulators should be turned on at all times. Instantaneously turning off the voltage regulator and turning on the current regulator, or vice versa, is not desirable because it tends to result in an instantaneous large current increase which causes current spikes. Furthermore, the power-on transient response is generally much worse than that of the set point change, and it is hard to realize seamless transition between two regulators. It is not practically feasible to turn on one regulator and turn off the other exactly at the same time, therefore we perform a make-before-break switching which has a period that both the regulators are turned on. However, different from ideal voltage and current sources, we should not simply tie up the outputs of the non-ideal voltage and current regulators because it may result in that current from the current source may flow into the voltage source. Practical voltage regulators do not allow reverse current which may cause hard failure in a power supply.

5.2.2.3 Circuit Design Principle

We propose a dual-mode power regulator circuit for the model-based PV emulator as shown in Fig. 5.11. It has adjustable voltage and current regulators whose outputs are tied together in parallel through two diodes. This parallel connection of diodes (or equivalently MOSFET-based lossless diodes) provides power hybridization

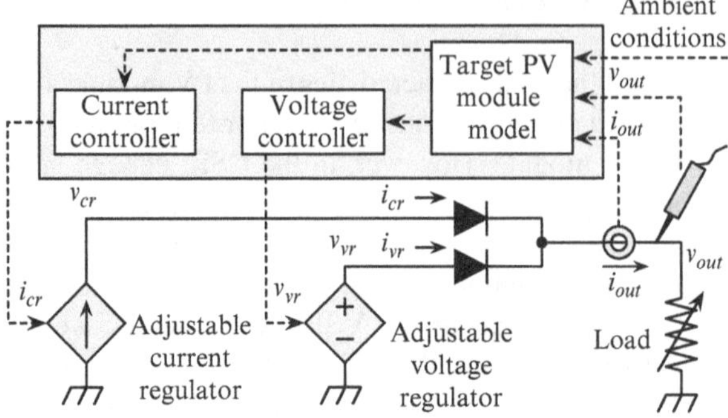

Fig. 5.11 Architecture of the proposed dual-mode power regulator circuit for PV emulators

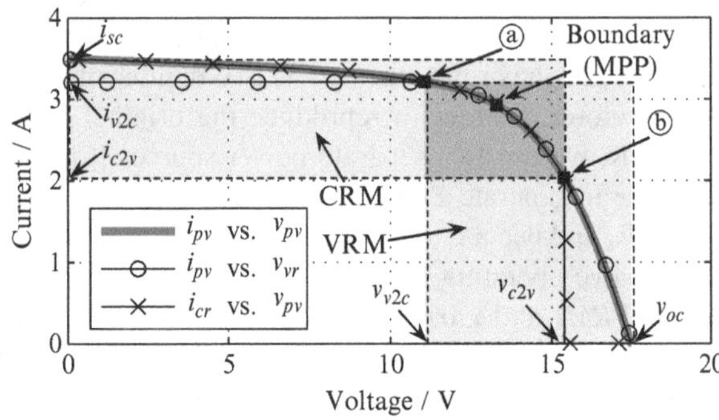

Fig. 5.12 I-V characteristics of the dual-mode power regulators compared with the target PV module I-V curve

preventing reverse current flow. This method has been used for power hot-swapping, and more recently, for hybridization of heterogeneous power sources [3], but has not yet been utilized for PV emulation.

The objective of the PV emulator is to make its output voltage V_{out} and output current I_{out} faithfully track V_{pv} and I_{pv}, which are derived from the PV module model. We switch the operation mode between the VRM and CRM near the boundary of the VSR and CSR. We first define a V-to-I mapping function and an I-to-V mapping function for the given I-V curve, based on (5.1)–(5.3). These functions translate V_{pv} to I_{pv} or vice versa, for given G and T.

Figure 5.12 shows three I-V curves of the target PV module model, the voltage regulator, and the current regulator. Here, the target PV module has a 17.5 V open-circuit voltage (V_{oc}) and 3.5 A short-circuit current (I_{sc}). The voltage V_{mpp} and current I_{mpp} at the MPP are 13.3 V and 2.9 A, respectively. We define V_{v2c} and I_{v2c} to be the voltage and current when the operating mode changes from the VRM to CRM, respectively, and V_{c2v} and I_{c2v} to be the voltage and current when the operating mode changes from the CRM to VRM, respectively. Both the operating points (V_{v2c}, I_{v2c}) and (V_{c2v}, I_{c2v}) are on the I-V curve, for the given G and T. These values are considered as the voltage or current limits in each operating mode. That is, the PV emulator generates the maximum output current of I_{v2c} in the VRM, and generates the maximum output voltage of V_{c2v} in the CRM.

We make VRM and CRM overlap across the boundary as shown in Fig. 5.12, and apply transition hysteresis. The transition hysteresis prevents frequent mode transitions between the VRM and CRM near the boundary. We make seamless transitions with the two-diode connection which allows make-before-break switching by blocking the reverse current. For example, when the operating mode switches from the VRM to CRM,

1. Keep the voltage regulator turned on and regulate output voltage V_{out} to V_{pv} while $I_{pv} < I_{v2c}$ (or equivalently, while $V_{pv} > V_{v2c}$).
2. If output current I_{out} increases and exceeds I_{v2c} (or equivalently, if V_{out} falls below V_{v2c}), turn on the current regulator (the transition point is annotated by ⓐ in Fig. 5.12).
3. When the current from the current regulator I_{cr} reaches I_{out}, turn off the voltage regulator and regulate output current I_{out} to I_{pv}.

The opposite is done when the operating mode switches from the CRM to VRM at ⓑ in Fig. 5.12; the voltage regulator is turned on first and current regulator is turned off later when the output current decreases. Both the voltage and current regulators maintain good controllability and thus high-quality output near the boundary, compared with the operating points far away from the boundary. Therefore, the proposed power hybridization and its control scheme guarantees smooth mode transition and guarantees superior output quality over the entire emulation range.

5.2.2.4 Dual-Mode Power Regulator Control

Figure 5.13 shows a control system block diagram of the proposed dual-mode power regulator circuit. It has two separate feedback control loops for the voltage and current regulators. The voltage and current regulators are coupled with each other such that their output V_{out} and I_{out} are located on the given I-V curve of the target PV module. The V-to-I and I-to-V mapping functions in Fig. 5.13, which are derived from the PV module model, handle the dual-mode operation.

The output behavior of the two-diode connection is such that its output voltage is either of the higher one between the two input voltages, and only the one with the higher voltage supplies the current. If two voltages are the same, the output current is the sum of the two input currents. The voltage and current control loop regulates voltage and current at different points. The current control loop regulates I_{cr}, and the voltage regulator control loop regulates V_{out} in order to compensate the voltage drop across the diode.

The hybridization controller ('Hybrid ctrl.' in Fig. 5.13) is in charge of such seamless transition of the operating mode between the VRM and CRM by the load demand. The controller takes six inputs: V_{c2v}, I_{v2c}, V_{out}, I_{out}, V_{vr}, and I_{cr}, and it generates two outputs: on/off signals for the voltage and current regulators. The operating mode transition point, V_{c2v} and I_{v2c}, are derived by the transition condition block ('Transition cond.' in Fig. 5.13). Figure 5.14 shows the functionality and behavior of the hybridization controller with a state machine. The state machine

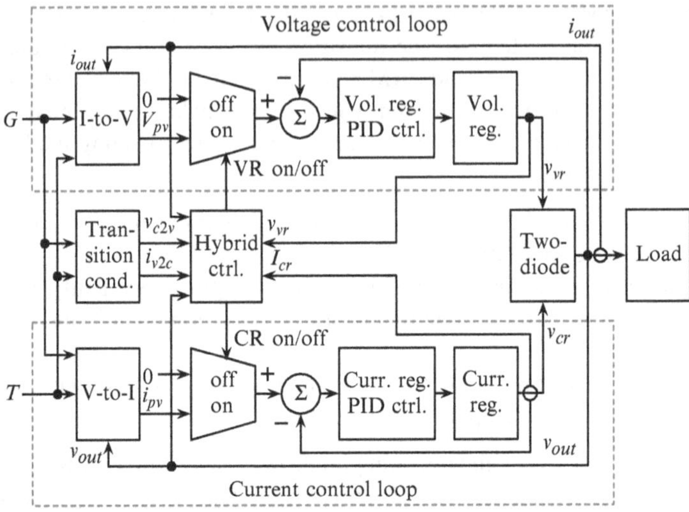

Fig. 5.13 Dual-mode power regulator control system block diagram

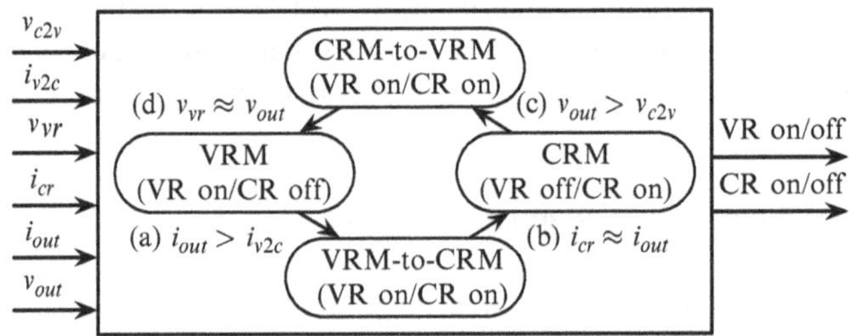

Fig. 5.14 State transitions and conditions of the hybridization controller (*VR*: voltage regulator, *CR*: current regulator)

has four states: the VRM, CRM, and two intermediate states between them. The output current, I_{out}, exceeding the limit of I_{v2c} makes transition from the VRM state. This results in that the current regulator is turned on and supplies current. If its output current I_{cr} becomes equal to I_{out}, that is, the voltage regulator does not supply current anymore, the state machine makes a transition to the CRM state and turns off the voltage regulator. The state machine makes transition from the CRM to VRM states in the same way. We consider the distance and position of the boundary between the VRM and CRM (ⓐ and ⓑ in Fig. 5.12) and determine the voltage/current limit V_{c2v} and I_{v2c}. The PV emulator will switch the mode too frequently if the gap is too narrow. On the other hand, the benefit from using the dual-mode power regulator is diminished if we expand the overlapping region and the gap becomes too wide.

5.2.2.5 Implementation

Figure 5.15 shows the implemented dual-mode regulator-based PV emulator board. Figure 5.16 shows the circuit schematic diagram of the proposed PV emulator. We use a low dropout (LDO) linear regulator LT1083 from Linear Technology as the voltage regulator. Switching regulators have a higher efficiency in general, but they are inherently subject to switching noise and voltage ripples even the load is constant. Since a PV module does not create any noisy ripples on its voltage and current, we use a linear regulator for the voltage source to eliminate them.

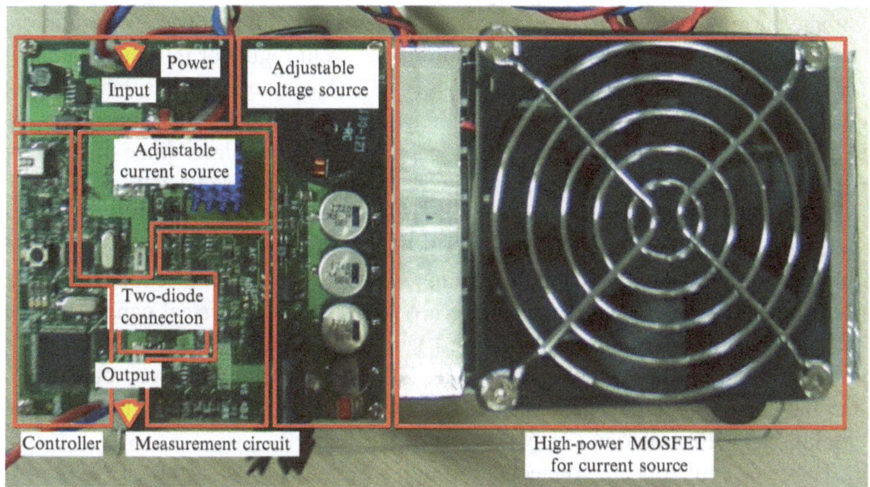

Fig. 5.15 Implemented dual-mode regulator-based PV emulator board

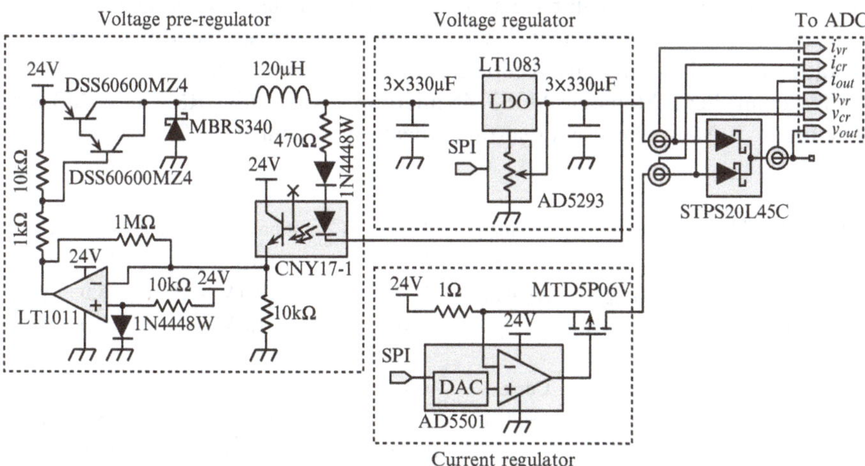

Fig. 5.16 Circuit schematic diagram of the PV emulator design

Power dissipated in a linear regulator is proportional to the dropout (voltage difference between the input and output). The dropout may be large in a wide output-adjustable regulator, and this results in a high heat dissipation. We implement a voltage pre-regulating circuit [8] for the input of the linear regulator to mitigate the heat dissipation. It automatically adjusts the input voltage to be higher that the output voltage by the required minimum dropout to minimize the heat dissipation, and allows a large amount of current from the linear regulator. The implemented voltage regulator is capable of supplying 1.2–21.8 V of output voltage with 12-bit resolution and up to 5.0 A of output current. However, the proposed dual-mode regulator architecture is not restricted to which type of regulator is used. We may use a high-efficiency switching regulator instead of the linear regulator as the voltage/current source to reduce its power dissipation for high-power PV emulation.

We implement a precision 10-bit resolution current regulator introduced in [10] for the current regulator also with the LT1083. The maximum available output current is 5.0 A up to 16.5 V, and it decreases as the output voltage increases up to 19.5 V. This does not limit the power capacity of the emulator since the current regulator generates a high current in the CRM where the voltage is low.

Due to the physical constraints of the components, the PV emulator's output voltage and current cannot span unlimitedly. Table 5.4 shows the voltage and current output ranges of the regulators. The range of operation of the regulators while performing the emulation is dependent not only on their physical capability, but also the I-V characteristic of emulating PV module. That is, the minimum output voltage that the voltage regulator generates is V_{v2c}, and the minimum output current that the current regulator generates is I_{c2v}, which vary depending on the target of emulation, as illustrated in Fig. 5.12.

The two regulators are connected through a Schottky diode array STPS20L45C from STMicroelectronics. It has a forward voltage drop in a range of 0.4–0.6 V varying depending on the temperature and forward current. It is described in the datasheet that the diode will have around 0.3 W of conduction loss at 1 A forward current due to the on-resistance. Therefore, we compensate the voltage drop and resulting power loss through a feedback control as discussed in Sect. 5.2.2.4.

The controller is Stellaris LM3S3748 microprocessor with ARM Corex-M3 core running at 50 MHz. We use a real-time operating system μC-OS II to implement

Table 5.4 Output specification of the implemented PV emulator. All values are measured at the output terminal

Regulator	Output	Conditions	Min.	Max.	Unit
Voltage regulator	Voltage	$I_{out} = 2.0$ A	1.2	21.8	V
	Current	$V_{out} = 10.0$ V	–	5.0	A
	Power	$V_{out} = 21.8$ V	–	109.0	W
		$I_{out} = 5.0$ A			
Current regulator	Voltage	$I_{out} = 2.0$ A	–	19.5	V
	Current	$V_{out} = 10.0$ V	0.0	5.0	A
	Power	$V_{out} = 16.5$ V	–	82.5	W
		$I_{out} = 5.0$ A			

the controller described in Sect. 5.2.2.4. The control task performs PID control at 1 kHz frequency. PID parameters are critical for fast response and stable operation of the PID controller to disturbances. We carefully set the PID parameters so that the voltage and current regulators have a fast response to the load variation and generate stable voltage and current output. We perform PID parameter tuning through extensive experiments and apply gain scheduling to achieve the best performance for the voltage and current regulators. The gain scheduling adjusts the PID parameters depending on the operation range in order to cope with the non-linear behavior. We divide the operation range into several subranges and tune the PID parameters for each subrange.

5.2.2.6 Experiments

We setup experimental environment as shown in Fig. 5.17. The implemented PV emulator is connected to the adjustable electronic load Kikusui PLZ334WL which can consume up to 300 W of power. The currents and voltages of the regulators and output are measured with a DAQ from the National Instruments. All the system including the PV emulator, electronic load, and DAQ is controlled by a customized automated tool.

PV Module I-V Characteristics

We first measure the output of the target PV module to compare the voltage and current variations in the VRM and CRM. Table 5.5 is the voltage and current variations of the target PV module caused by load power variations in the VRM and CRM. Refer to Fig. 5.10 which graphically illustrates the voltage and current

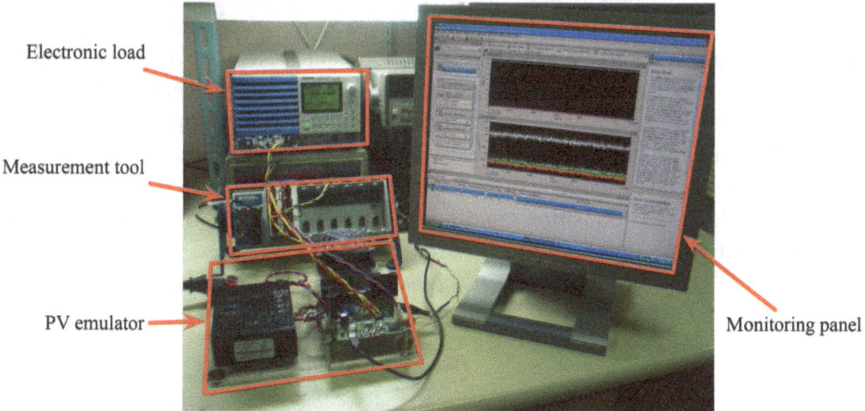

Fig. 5.17 Experimental setup

Table 5.5 Voltage and current variations by the load power variation of 40–60 % of the maximum power at 1,000 and 500 W/m^2

G (W/m^2)	Region	Output	Marker in Fig. 5.10	Range	Δ_{norm} (%)
1,000	VSR	Voltage	②	16.2–16.7 V	2.9
		Current	③	0.93–1.44 A	14.6
	CSR	Voltage	④	4.5–6.9 V	13.7
		Current	⑤	3.38–3.43 A	1.4
500	VSR	Voltage	②	15.7–16.2 V	2.9
		Current	③	0.47–0.72 A	14.3
	CSR	Voltage	④	4.4–6.7 V	13.5
		Current	⑤	1.69–1.71 A	1.1

variations caused by the power variation. We measure the voltage variation (② and ④) and current variation (③ and ⑤) while changing the load power (①) in a range of 40–60 % of the maximum power for the given irradiance. Table 5.5 presents the range of voltage and current variations in each region and the normalized value Δ_{norm} to the maximum voltage and maximum current, which are V_{oc} and I_{sc}, respectively. This definitely shows that using an appropriate power regulator results in only 1–3 % output variations, otherwise it suffers from 13 to 15 % output variations. This phenomenon becomes more clear when the operating point is near the maximum voltage (open-circuit) or maximum current (short-circuit). This confirms the dual characteristic of the PV module and shows the potential benefits of using two different regulators for PV emulation.

Matlab/Simulink Simulation

We first validate the functionality of the proposed circuit and the control method with Matalb/Simulink simulation. Through the simulation, we show that the proposed circuit well performs the PV emulation, and present the resulting voltage, current, and power behavior. We use adjustable voltage and current regulator models, diode models, and resistive load models from the Matlab/Simulink Simscape library. Without loss of generality, we use a pre-measured PV module I-V curve at a given irradiance level and temperature for demonstration purpose.

First, we define the transition conditions in Fig. 5.14. We define I_{v2c} in (a) and V_{c2v} in (c) as follows:

$$I_{v2c} = (V_{mpp} + V_{oc})/2, \quad (5.10)$$

$$V_{c2v} = (I_{mpp} + I_{sc})/2. \quad (5.11)$$

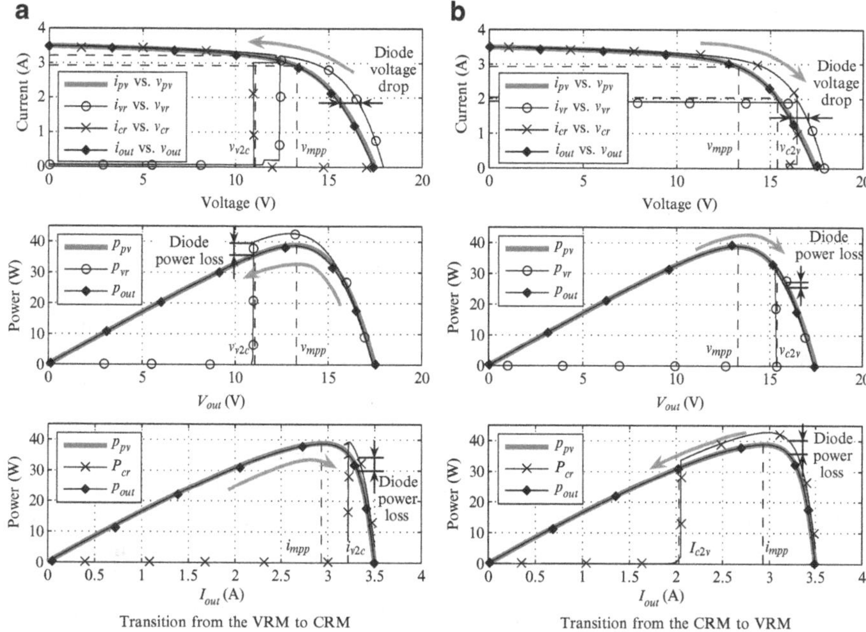

Fig. 5.18 I-V, P-V, and P-I curves of the voltage regulator, current regulator, and emulator output, compare with the PV module model while transiting (**a**) from the VRM to CRM, and (**b**) from the CRM to VRM

We consider that $V_{vr} \approx V_{out}$ in (b) and $I_{cr} \approx I_{out}$ in (d) when the error is less than 0.1 %. These conditions are empirically determined based on the observation on the I-V curve of the target PV module.

Figure 5.18 shows the I-V curves, P-V curves, and P-I curves of the PV module model, voltage regulator, current regulator, and emulator output. Figure 5.18a, b show the two cases such that the operating point changes in two different directions, respectively, which are denoted by a gray arrow. It demonstrates the hybridization controller described in Sect. 5.2.2.4 is functioning as expected. The operating mode transition occurs when the output voltage reaches to V_{v2c} in the VRM in Fig. 5.18a or when the output current reaches to I_{c2v} in the CRM in Fig. 5.18b. We see from the I-V curves that the voltage output of the voltage and current regulators is higher than V_{pv} (shifted to higher than V_{pv}) as a result of the feedback control to compensate the diode forward bias voltage drop of 0.6 V and the diode on-resistance of 0.3 Ω. In spite of the steep I-V curve at the transition, the resultant PV emulator output I-V characteristic well matches with that of the PV module model thanks to the use of two-diode hybridization circuit and control.

I-V Characteristics

Now that we have validated the functionality through simulation, we present the measurement results obtained from the physical experiments. We show the output quality of the proposed PV emulator compared with two conventional voltage or current regulator-based PV emulators. We turn off the current or voltage regulator to generate the output of voltage regulator-based PV emulator or current regulator-based emulator, respectively.

Figure 5.19a–c show the I-V curves measured from the three PV emulators based on the voltage regulator, current regulator, and dual-mode regulator, respectively. We measure the voltage and current while changing the load from the with period of 30 s. The load changes between zero (open circuit) to the value that makes the output 2 V, which is close to the minimum output voltage described in Table 5.4. In each figure, the solid line denotes the desired output I-V characteristic according to the PV module model, and the markers denote the measured points.

In Fig. 5.19a, b, it is definite that the output of the regulators are not as expected in some operating range. More specifically, the voltage regulator fails to generate the desired voltage in the CSR as annotated by ⓐ. Similarly, the current regulator fails to generate the desired current in the VSR as annotated by ⓑ. This is not because the regulators are not capable of generating high power in those regions, but because the output voltage or current variation is too rapid for the regulator to follow the output change. The power output capability of the regulators is enough to generate the voltage and current on the target I-V curves as presented in Table 5.4. For example, the voltage regulator can stably supply up to 5 A at 10 V, but the I-V curve of 500 W/m^2 in Fig. 5.19a shows unstable output at low current and low voltage range below 1.8 A and 7 V. In contrast, the output of the dual-mode regulator is in a good quality as shown in Fig. 5.19c. The output in the VSR is as good as that of the voltage regulator-based emulator, and the output in the CSR is as good as that of the current regulator-based emulator.

In practice, a load device connected to a PV module is not only a resistance-mode load, but may be a voltage-mode, current-mode, or combination of them. Therefore, we show that the proposed dual-mode PV emulator shows a good stability over all operating range not only for the resistance-mode load as shown in Fig. 5.19c, but also for voltage- and current-mode loads. Figure 5.20a, b show I-V curves when a voltage- and current-mode load is applied, respectively. We can see that both I-V curves for the voltage- and current-mode loads exhibit good consistency with the reference I-V curve.

Mode Transitions

Figure 5.21 shows the voltage and current output variations of the voltage regulator, current regulator, and emulator output when the load changes. This is a representa-

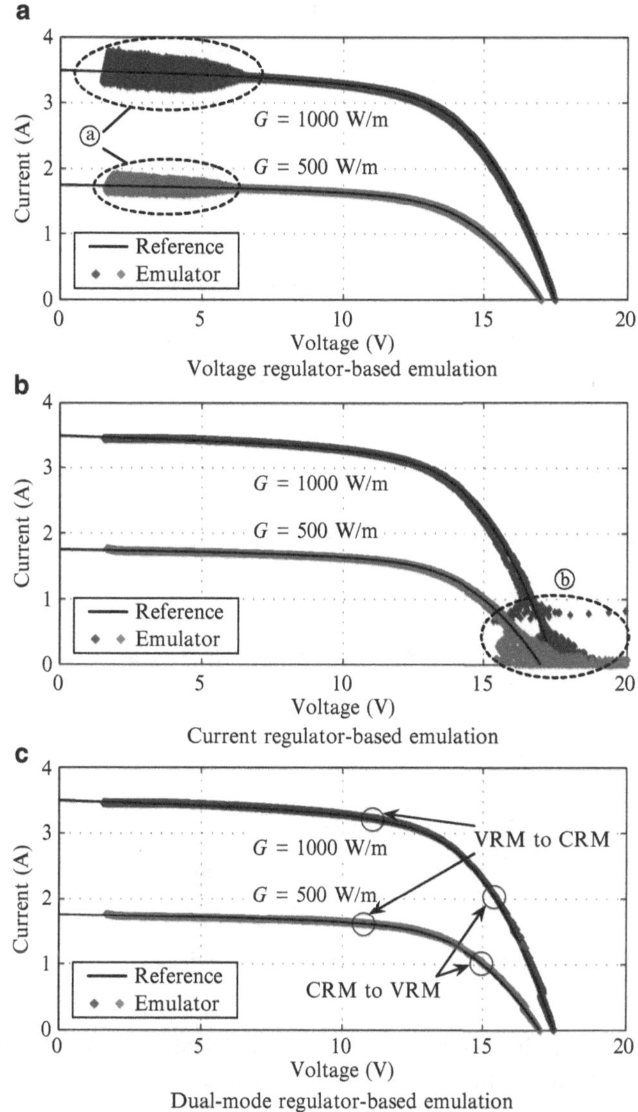

Fig. 5.19 I-V curves of PV emulation based on three regulators: (**a**) voltage regulator, (**b**) current regulator, and (**c**) dual-mode regulator

tion of the data in Fig. 5.19c in a time axis. We apply a variable load starting from a zero load (open-circuit) to a very low resistance load (near short-circuit), and back to a zero load.

It starts in the VRM state because the load is zero. Current draw increases, and at t_1, the output current I_{out} reaches I_{v2c} which is the current limit of the voltage

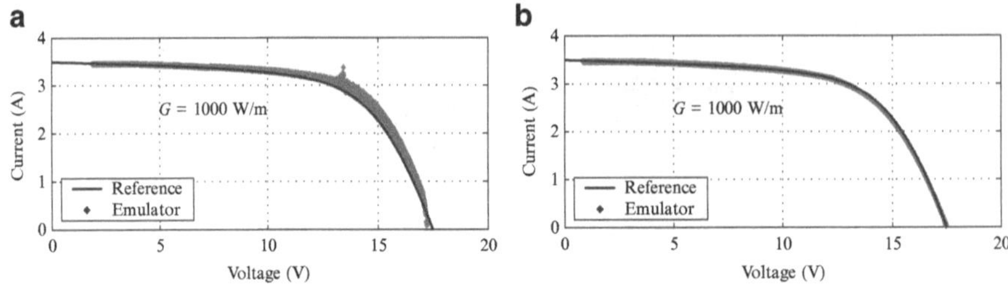

Fig. 5.20 I-V curves of PV emulation for (**a**) a voltage-mode load, and (**b**) a current-mode load

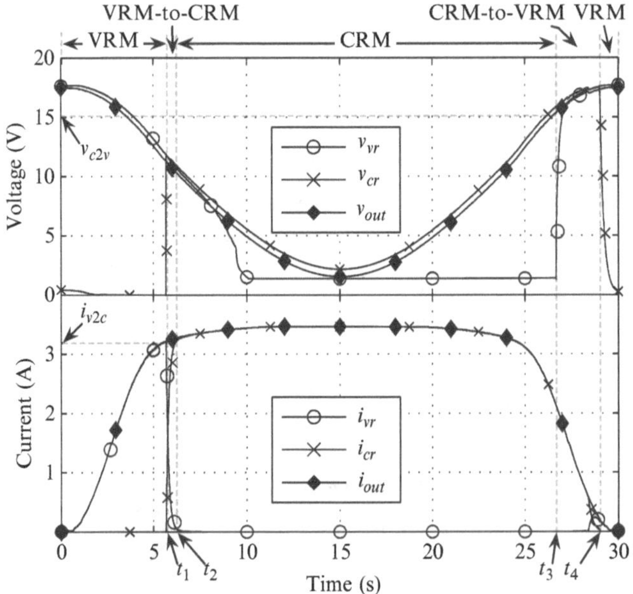

Fig. 5.21 Output of the voltage and current regulators for varying load

regulator. The operating mode changes from the VRM to VRM-to-CRM state, in which I_{cr} gradually increases, and I_{vr} gradually decreases. Finally, I_{cr} reaches to I_{out} at t_2, and the voltage regulator is turned off, by entering the CRM state. We see that the operating mode transition is seamlessly performed, and the output voltage V_{out} and current I_{out} are stably maintained during the transition. The hybrid controller performs the opposite when the operation mode changes from the CRM to VRM. The output voltage V_{out} increases and current I_{out} decreases until it reaches V_{c2v} at t_3. The controller enters the CRM-to-VRM state and the voltage regulator is turned on. When the output voltage of the voltage regulator reaches V_{out} at t_4, the current regulator is turned off by entering the VRM state.

Next, we show the PV emulator's transient response to a step load change. A PV emulator should be able to change its operating point rapidly when the load changes.

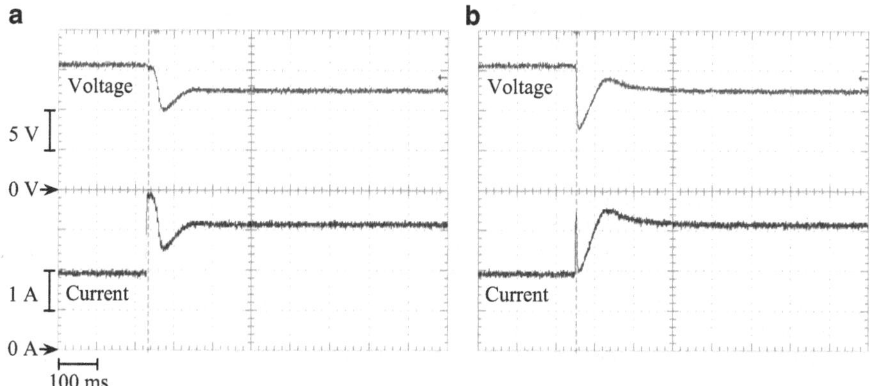

Fig. 5.22 Voltage and current variance for step change of load resistance from 8.3 to 4.0 Ω. (**a**) Voltage regulator. (**b**) Current regulator

Due to the control hysteresis, the same operating point can be regulated by either voltage or current regulator, especially near the MPP. Therefore, we apply a step change of the load resistance and observe the voltage and current variations [2]. The step response of the voltage regulator and current regulator is shown in Fig. 5.22a, b, respectively. The MPP is at 2.9 and 13.3 A, which corresponds to 4.5 Ω. The load resistance is first 8.3 Ω and the PV emulator output is 1.9 A at 15.6 V. We decrease the resistance to 4.0 Ω and the operating point changes to 3.1 A at 12.4 V. Both the voltage and current converters to the desired operating point within about 100 ms.

References

1. Alonso-García M, Ruiz J, Chenlo F (2006) Experimental study of mismatch and shading effects in the I-V characteristic of a photovoltaic module. Solar Energy Materials and Solar Cells
2. Durán E, Andújar J, Segura F, Barragán A (2011) A high-flexibility DC load for fuel cell and solar arrays power sources based on DC–DC converters. Applied Energy 88(5):1690–1702
3. Kim Y, Shin D, Seo J, Chang N, Cho H, Kim Y, Yoon S (2010a) System integration of a portable direct methanol fuel cell and a battery hybrid. International Journal of Hydrogen Energy 35(11):5621–5637
4. Kim Y, Wang Y, Chang N, Pedram M (2010b) Maximum power transfer tracking for a photovoltaic-supercapacitor energy system. In: Proceedings of the ACM/IEEE International Symposium on Low-Power Electronics and Design (ISLPED), pp 307–312
5. Kim Y, Lee W, Pedram M, Chang N (2013) Dual-mode power regulator for photovoltaic module emulation. Applied Energy 101(0):730–739
6. Koran A, Sano K, Kim RY, Lai JS (2010) Design of a photovoltaic simulator with a novel reference signal generator and two-stage LC output filter. IEEE Transactions on Power Electronics 25(5):1331–1338
7. Lee W, Kim Y, Wang Y, Chang N, Pedram M, Han S (2011) Versatile high-fidelity photovoltaic module emulation system. In: Proceedings of the IEEE/ACM International Symposium on Low-power Electronics and Design, pp 91–96

8. Linear Technology (1994) LT1083: 7.5A, 5A, 3A low dropout positive adjustable regulators
9. Lu C, Raghunathan V, Roy K (2010) Maximum power point considerations in micro-scale solar energy harvesting systems. In: Proceedings of IEEE International Symposium on Circuits and Systems (ISCAS), pp 273–276
10. Neubauer J (2004) Precision current source is software-programmable. EDN p 72
11. Sanchis P, Echeverria I, Ursua A, Alonso O, Gubia E, Marroyo L (2003) Electronic converter for the analysis of photovoltaic arrays and inverters. In: Proceedings of Power Electronics Specialist Conference (PESC), vol 4, pp 1748–1753
12. Sera D, Teodorescu R, Rodriguez P (2007) PV panel model based on datasheet values. In: Proceedings of the IEEE International Symposium on Industrial Electronics (ISIE), pp 2392–2396, DOI 10.1109/ISIE.2007.4374981
13. Villalva M, Gazoli J, Filho E (2009) Comprehensive approach to modeling and simulation of photovoltaic arrays. IEEE Transactions on Power Electronics 24(5):1198–1208
14. Xiao W, Dunford W, Palmer P, Capel A (2007) Regulation of photovoltaic voltage. IEEE Transactions on Industrial Electronics 54(3):1365–1374

Chapter 6
Implementation and Application

In this section, we present two experimental results. We first present a simulation of the proposed optimization technique applied to an EV. The EV presented in the simulation is comprised of a PV module, a regenerative brake, and a battery-supercapacitor HEES system. We apply the proposed design optimization technique and power management scheme in the EV and show an improved energy efficiency. In the second part of the experiment, we present an implementation of a HEES system composed of lead-acid batteries, Li-ion batteries, and supercapacitors. We describe the details of design and implementation of hardware and software, and present various measurement results. The content of this chapter is in part based on [2].

6.1 EV Application

We present an example that the proposed HEES system design and operation optimization methods applied to an EV. The HEES application is easy to see the benefit of the high energy efficiency by directly converting into fuel cost reduction. Figure 6.1 shows the components of an EV. It is equipped with both a conventional combustion engine and an electrical traction motor. The regenerative brake produces electrical energy when braking. PV modules installed on the roof and bonnet also produces energy from the solar irradiance. The energy generated by the regenerative brake and PV modules are stored in the energy storage, and used later to operate the traction motor for acceleration.

The baseline for comparison is a battery-only EES system without reconfiguration. The battery has 375 V output voltage and 53 kWh energy capacity. The PV modules generates 100 W maximum power in total and perform MPPT to generate maximum power from the PV modules. All the electrical devices, traction motor, regenerative brake, energy storage, and PV modules, are connected through

Y. Kim, N. Chang, *Design and Management of Energy-Efficient Hybrid Electrical Energy Storage Systems*, DOI 10.1007/978-3-319-07281-4_6

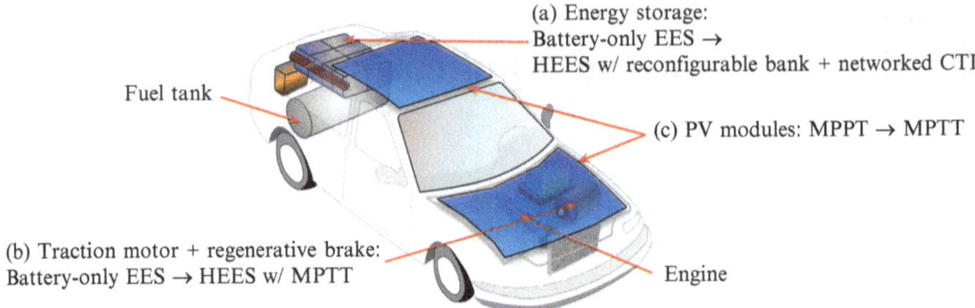

Fig. 6.1 An EV with PV modules. Energy storage, motor/brake, and solar modules are applied the proposed optimization methods

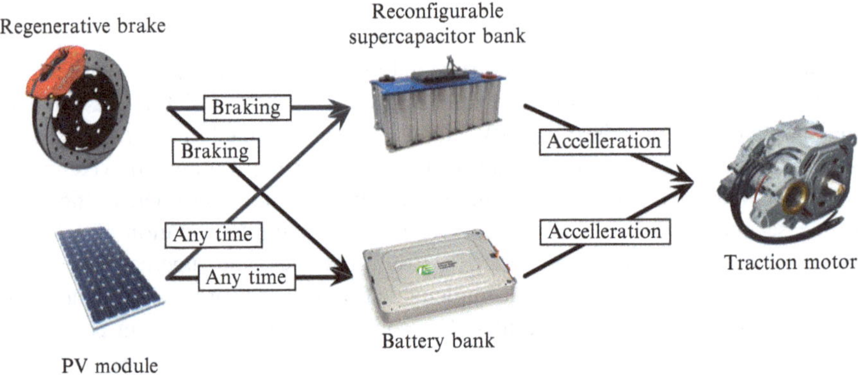

Fig. 6.2 Power flow among components

a shared bus CTI. We improve the EV by applying the proposed HEES system design and operation methods. We make HEES system by adding a supercapacitor bank composed of 140 cells which has 1,200 F capacity each. Figure 6.2 illustrates the power flow among the components. The following sections introduce how the proposed optimization methods improves the energy efficiency in the EV in each component by comparing with the baseline EV.

6.1.1 Regenerative Brake

We increase energy recovery form the regenerative brake by hybrid use of the supercapacitor bank with the battery bank. We also apply the MPTT to further increase the energy efficiency. Power density of batteries are not so high that they are not suitable for handling intermittent high power demand. The regenerative brake produces a very large amount of power for a short duration, and so a battery-only EES system suffers from low energy efficiency due to the poor rate-capability.

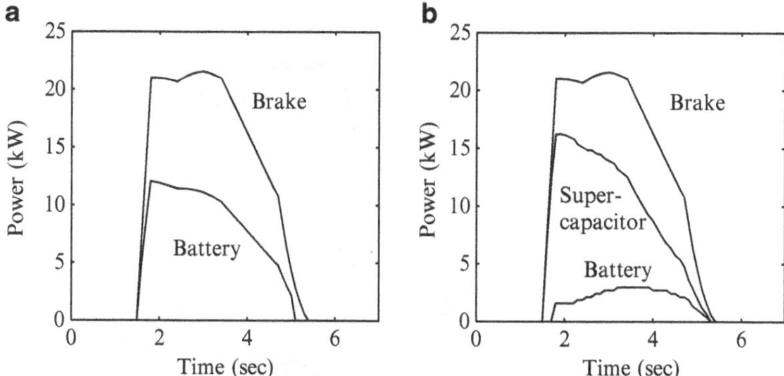

Fig. 6.3 Power output from the brake and power input to the battery and supercapacitor banks while braking. (**a**) Battery-only EES. (**b**) Battery-supercapacitor EES /w MPTT current distribution

Hybrid use of supercapacitor bank greatly improves the energy efficiency, and thus recovers more energy. While the amount of current charged into the battery in a EES system is directly determined by the energy generation from the regenerative brake, having a battery-supercapacitor HEES system incurs charge allocation problem to determine the amount of current to each bank.

We determine the current distribution between the battery and supercapacitor based on the charge allocation policy proposed in [1]. This policy determines the current distribution so that the amount of total recovered energy is maximized. The energy is not evenly distributed between the battery and supercapacitor because they have different rate capability. Also, the current distribution changes over time depending on the input from the regenerative brake changes. We also apply the proposed MPTT operational method to the power converters in the battery bank and supercapacitor bank. Figure 6.3 shows the power output from the regenerative brake and power input to the battery and supercapacitor banks. The EV reduces its velocity from 70 to 0 km/h for about 4 s. The battery voltage is 375 V and initial supercapacitor voltage is 252 V. This is a case that the power conversion efficiency between the brake and supercapacitor is high thanks to similar voltage levels. Amount of recovered energy with the battery-supercapacitor HEES system is 49 % higher than the battery-only EES system. The improvement varies depending on the supercapacitor voltage. Assuming uniform probability of supercapacitor voltage, the additional energy recovery per one braking operation is about 3.1 Wh on average.

6.1.2 PV Modules

We increase power generation by applying the proposed MPTT method to the 100 W the PV modules. Figure 6.4 shows that energy harvesting increases by the proposed MPTT operation compared with the conventional MPPT operation.

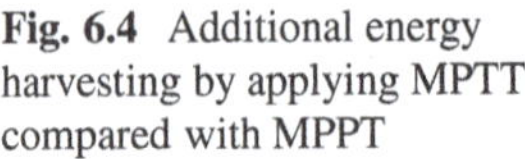

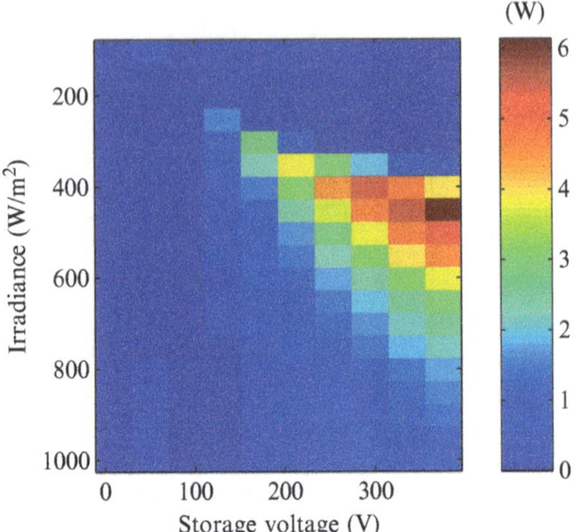

Fig. 6.4 Additional energy harvesting by applying MPTT compared with MPPT

The MPTT operation maximizes the current output from the converter. Assuming uniform probability of supercapacitor voltage and irradiance level, the additional energy harvesting is 0.9 Wh per hour on average.

6.1.3 EES Bank Reconfiguration and Networked CTI

We demonstrate energy efficiency improvement by the proposed EES bank reconfiguration and networked CTI architecture. We reconfigure the supercapacitor bank so that the power charged into the supercapacitor bank is maximized when it is charged by the regenerative brake and PV modules. We make a sequence of charge transfers tasks to the network CTI design problem based on the improvement described in Sects. 6.1.1 and 6.1.2. The vehicle accelerates using the energy in the supercapacitor bank firstly. It uses energy in the battery bank if the energy is not enough. Both the acceleration and braking take 4 s each. Between the acceleration and braking, the vehicle cruises (maintains the constant velocity) for 1 min using the combustion engine only. We assume that the solar irradiance level is 750 W/m^2.

Figure 6.5 shows the CTI architecture and connectivity of the baseline EV and the proposed EV. The baseline EV uses a single shared-bus CTI as shown in Fig. 6.5a to connect four nodes. The charge transfer efficiency in this shared-bus architecture is 46 %. The proposed EV has five nodes (a supercapacitor bank is added) and they are connected though a networked CTI, as shown in Fig. 6.5b. The charge transfer efficiency in this networked CTI architecture is improved 51 %. The charge transfer efficiency slightly further increases to 53 % if we apply EES bank reconfiguration for the supercapacitor bank so that it maintains the best configuration for the PV and reintegrative brake.

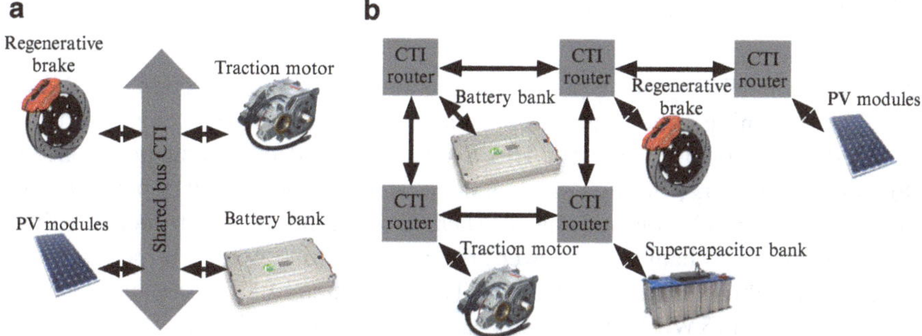

Fig. 6.5 CTI architecture and connectivity of the baseline shared-bus CTI and proposed networked CTI. (**a**) Battery-only homogeneous EES with shared bus CTI (baseline). (**b**) Battery-supercapacitor HEES with networked CTI (connectivity as shown)

Fig. 6.6 Energy recovery/harvesting gain per cycle. (**a**) MPTT regenerative braking with networked CTI. (**b**) MPTT regenerative braking with bank reconfiguration. (**c**) MPTT solar energy harvesting with networked CTI. (**d**) MPTT solar energy harvesting with bank reconfiguration

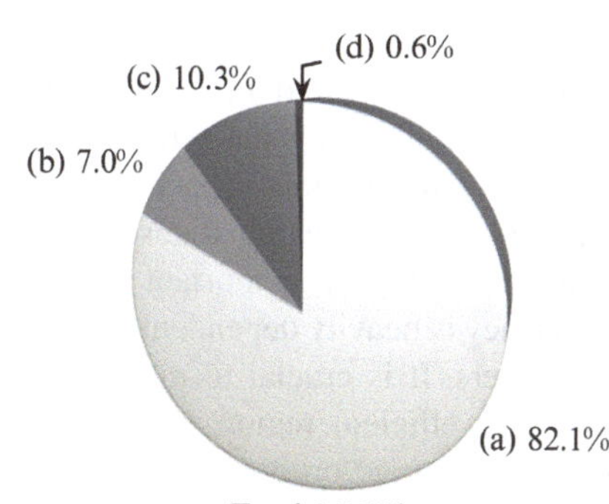

6.1.4 *Overall Improvement and Cost Analysis*

Figure 6.6 shows contributions of each design and optimization methods to the energy efficiency improvement. The total energy gain per cycle is 3.7 Wh with the usage scenario described in Sect. 6.1.3. Majority of energy gain, which is as high as 82.1 %, comes from the MPTT operation of the regenerative brake with the networked CTI battery-supercapacitor HEES system. Dynamic supercapacitor bank reconfiguration achieves additional 7.0 % more energy gain from the regenerative brake. Energy gain from the PV modules is improved by 10.3 and 0.6 % by the networked CTI and supercapacitor bank reconfiguration, respectively. The energy gain from the regenerative brake becomes significant as the acceleration and braking cycles become more frequent.

6.2 HESS Prototype Implementation

We discuss the HEES system prototype implementation in this section. The HEES prototype is aiming at household applications though the design framework is not bounded to a particular scale. The HEES system is connected to the power grid to store electrical energy during non-peak hours and becomes auxiliary power source to mitigate the peak power demand. There are five nodes which includes three heterogeneous EES banks, one power grid input, and one AC power outlet, which are connected through a shared-bus CTI. There are three EES banks such as 6.5 Wh supercapacitor, 115 Wh Li-ion battery, and 163 Wh lead-acid battery banks. The HEES system provides a high degree of freedom to transfer energy between the nodes at a high efficiency by continuous update of the current and voltage of each EES bank and CTI.

The HEES system mandates elaborate management policies because the energy flow in the HEES is much more complicated than conventional heterogeneous type EES systems. It may suffer from a very low energy efficiency unless we perform intelligent management with consideration of characterizations of EES elements (IR loss, rate capacity effect, leakage, etc.). While most previous electric energy storage research focused on storage elements, we emphasize that power converters play an important role in the HEES system. HEES system energy efficiency is heavily dependent on the input and output voltage and current of power converters. It is crucial to maintain the power converter operating point close to the most efficient region. The charger board proposed this dissertation consists of a wide-range programmable bidirectional charger and a microcontroller with Control Area Network (CAN) interface. We have shown that proper determination of voltages and currents of each EES bank and CTI increases energy efficiency of the HEES system and introduced system-level management policies including charge allocation [12], charge replacement [13], and charge migration [9, 10]. We implement the management algorithms in the main controller that communicates with the charger boards through CAN.

6.2.1 *Design Specifications*

We describe the design specification of the HEES prototype implementation. We design a HEES system for load leveling and peak shaving of residential electricity usage. The specification of the HEES system in this section is for the proof-of-concept purpose. That is, the architecture and control method are highly scalable and flexible so that they can be applied to smaller and larger scale HEES systems with various types of EES elements, power sources, and load devices.

6.2.1.1 Power Input and Output

The input to the HEES system is 120 V AC, and the output is 120 V AC. The proposed HEES system is transparent to both the power grid and load devices; it is seen as an ordinary AC-powered appliance to the power grid, while it is a AC power outlet to the load devices.

6.2.1.2 Power and Energy Capacity

We set the power capacity of the HEES system to 300 W. This is reasonable power capacity for load leveling and peak shaving purposes considering that the average residential electric usage ranges 0.4–1 kW over time [11]. All the EES banks are designed to meet this power capacity at least. We set the total energy capacity of the HEES system as small as 300 Wh for shorten time of charging and discharging the system for experimental purpose. Nevertheless, the energy capacity may be easily increased by extending the EES array or adding more EES banks.

6.2.1.3 Voltage and Current Ratings

Maximum power efficiency is available with proper operational range of voltage and current. Typically, power converters with hundreds of kW power capacity are designed with a DC voltage range of 12–48 V and maximum current of 5–20 A. Commercial DC–DC converters with symmetric input and output voltage ranges have such voltage and current ratings typically. AC–DC rectifiers and DC–AC inverters have similar voltage and current ratings for the DC side. We set the maximum current to 10 A for each power converter. We design the power converters accordingly, and the voltage range covers a much wider range of 6–36 V for voltage variation of the supercapacitors and CTI.

6.2.1.4 EES Elements

We use three types of EES elements in the proposed HEES system prototype: supercapacitor, Li-ion battery and lead-acid battery. The supercapacitor has advantages in power capacity cycle life, and cycle efficiency, while the lead-acid battery has advantages in cost. The Li-ion battery has moderately good characteristics all round, except for the cost. We do not consider other types of EES elements such as kinetic or thermal storages, but the proposed HEES system does not restrict types of EES elements fundamentally.

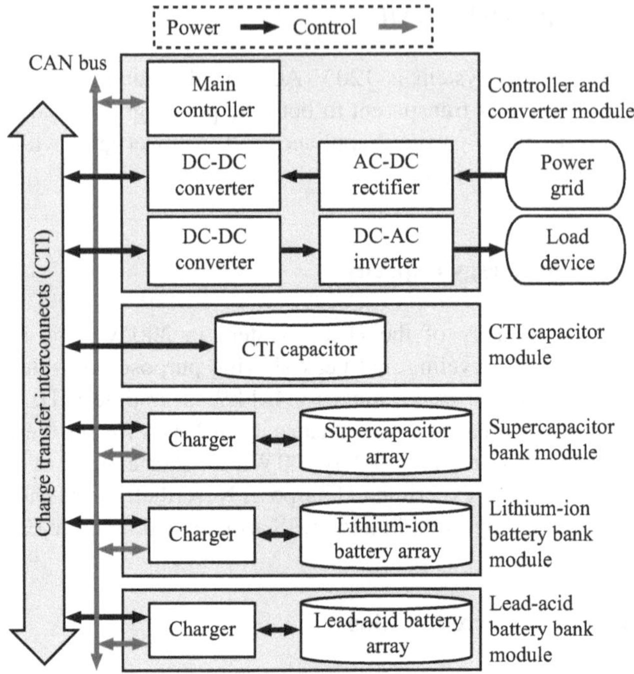

Fig. 6.7 Architecture of the proposed HEES system

6.2.2 *Implementation*

We discuss the implementation of the HEES system prototype in this section. We especially focus on showing that the implementation satisfies the specifications of Sect. 6.2.1 with consideration on the design goals of Sect. 3.1. We achieve the goal of the high modularity of the HEES system by modular implementation of the HEES system. The modular design makes it easier to develop and modify EES bank individually and manipulate the system-level configurations. Figure 6.7 shows modules of the proposed HEES system composed of three EES bank modules, controller and converter module, and CTI capacitor module. The following subsections describe design and implementation of the modules and their subcomponents, justifying the design goals and specifications.

6.2.2.1 Bank Module

The proposed HEES system has three EES banks: a supercapacitor bank, a Li-ion battery bank, and a lead-acid battery bank. Figure 6.8a shows the supercapacitor bank module viewed from the front and back assembled in a 19-in. rack case. There is an EES array composed of 18 supercapacitors inside the rack case, and

a

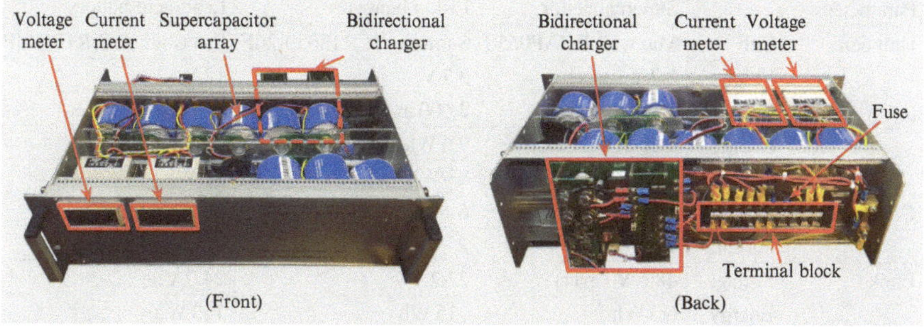

b **c**

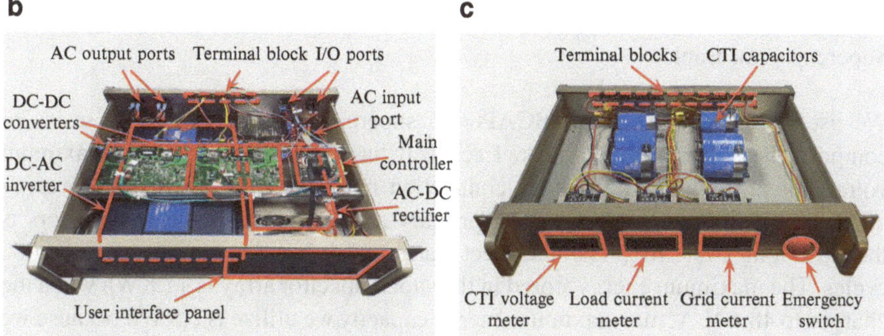

Fig. 6.8 Modules of the proposed HEES system. (**a**) Supercapacitor bank (front and back). (**b**) Controller and converter module. (**c**) CTI capacitor module

a bidirectional charger is attached to the back of the module together with terminal blocks. The voltage and current meters on the front panel display the voltage and current of the EES array. We put a fuse to prevent excess current flow between the bank and CTI that may cause damage to the array or charger. The Li-ion battery module bank and the lead-acid battery bank module also have a similar structure. Composition each EES bank module is summarized in Table 6.1.

The EES bank implementation in a 19-in. rack case as a module is for scalability and modularity of the system. Design of each EES bank is independent to each other. All the EES banks have the uniform interface regardless the EES elements. It has a DC power port to the CTI, a communication network port, and a power input port for the bidirectional charger. Therefore, changing the configuration of an EES bank module or adding/removing a module is simple and has limited influence to the whole system.

Table 6.1 Composition of EES banks

Parameter		Supercapacitor	Li-ion battery	Lead-acid battery
Unit cell	Cell	Maxwell BCAP0650	Samsung ICR18650-26F	Panasonic LC-R123R4P
	Voltage	2.7 V (max)	3.7 V	12 V
	Capacity	650 F	2,600 mAh	3,400 mAh
	Energy	0.66 Wh	9.6 Wh	40.8 Wh
	Cost	$69/Wh	$3/Wh	$1/Wh
Configuration	Series	18	6	2
	Parallel	1	2	2
Bank	Voltage	48.6 V (max)	22.2 V	24.0 V
	Energy	6.3 Wh	115 Wh	163 Wh

Supercapacitor Bank

We use 18 cells of Maxwell BCAP0650 supercapacitor connected in series to compose the supercapacitor bank. Each cell has 650 F capacity with maximum voltage of 2.7 V, and so 18-series connection makes a 36 F supercapacitor array of maximum voltage of 48.6 V. The primary advantage of the supercapacitors is the extra long cycle life. The datasheet states that it has a cycle life of 1,000,000 cycles. The maximum energy stored in the supercapacitor array is 11.8 Wh when the charged to 48.6 V. Actual maximum energy capacity we utilize is 6.3 Wh because we limit the terminal voltage of the supercapacitor array within 6–36 V as specified in Sect. 6.2.1. Cost per energy of the supercapacitor is about $69/Wh based on the retail price for purchasing single cell, which is much more costly than that of batteries by an order of magnitude.

Li-Ion Battery Bank

We compose the Li-ion battery bank with 12 cells of Samsung ICR18650-26F [6]. Each cell has 2,600 mAh capacity and nominal voltage of 3.7 V. We arrange the 12 cells into a 6×2 array which give us 22.2 V nominal voltage and 115 Wh energy capacity. The number of series connection is determined to make the nominal voltage be placed in the middle of voltage range of 6–36 V in order to mitigate power efficiency degradation which may be caused by voltage difference between the battery array and CTI. The total energy capacity is chosen to be small so as to shorten the time for experiments as we mentioned in Sect. 6.2.1. In fact, the rack case has a plenty of room for additional batteries.

Lead-Acid Battery Bank

The lead-acid battery bank consists of four Panasonic LC-R123R4P batteries [5]. Each battery is composed of six series cells which make nominal voltage of 12.0 V. We make a 2 × 2 array to obtain a terminal voltage of 24.0 V. The lead-acid battery bank has the largest energy capacity thanks to its low cost, which is only $1/Wh based on the retail price for purchasing a single battery. The total energy capacity of the lead-acid battery bank is also set small for fast experiments.

6.2.2.2 Controller and Converter Module

The controller and converter module performs functionalities of system supervision and AC conversions (AC-to-DC and DC-to-AC conversions). The system supervision is conducted by the main controller which is implemented with a Texas Instruments LM3S2965 microcontroller running at 50 MHz. The microcontroller runs Micrium μC-OS II real-time operating system (RTOS) and monitoring, control, communication tasks on top of it. The microcontroller provides an enough computing power for simple control policies with real-time control loops, as well as connectivity to the PC with much more powerful computing capability for more elaborated control policies.

The main controller communicates with subsystems of the HEES systems at 1 Hz–1 kHz frequency depending on the data. It with the bidirectional chargers through the CAN bus for monitoring and control of the EES banks. We adopt CAN bus for the communication network within the HEES system. The CAN bus is a widely-used industrial standard for system control, and supports multi-master communication with up to 1 Mbit/s bit rate within tens of meters range. It provides a high scalability for multiple subsystems to be connected through the network that is hard to achieve with other types of communication networks. The data transfer between an EES bank and the main controller is less than 10 kbyte/s (1 byte = 8 bits), and so 1 Mbit/s bit rate is more than enough for the HEES system with small number of EES banks. The main controller also communicates with the user interface panel which provides useful information such as voltage and current of each EES bank, CTI, and power grid. In addition, the main controller can be connected to a PC so that the user can monitor and control the HEES system manually. We discuss the software design for the main controller and the bidirectional charger in Sect. 6.2.2.4 in more detail.

The controller and converter module also performs both AC-to-DC conversion from the power grid to the CTI and DC-to-AC conversion from the CTI to the AC load devices. Since our focus is on the system design of the HEES system, not power converter circuit design, we use commercial high-efficiency AC converters rather than designing custom AC converters. We use the Mean Well SE-600 600 W AC–DC rectifier. We need an adjustable output voltage in order to dynamically change the CTI voltage, but it generates a 36 V fixed output voltage. Therefore, between the AC–DC rectifier and the CTI, we implement a DC–DC converter which accepts 36 V input voltage and generates an adjustable wide output voltage for the CTI. The power converter connection is shown in Fig. 6.7. The output voltage level of the DC–DC converter is controlled by the main controller.

We use Samlex PST-100S-24A 1,000 W pure sine wave DC–AC inverter. In contrast to the AC–DC rectifier, the DC–AC inverter is required to accept wide input voltage of 6–36 V of the CTI, but typical commercial inverters designed for batteries do not support such a wide input range. The PST-100S-24A is operational only for 21.4–33 V input, and the nominal input voltage is 24 V. Therefore, similar to the AC–DC rectifier, we implement a DC–DC converter which accepts 6–36 V input voltage and generates 24 V fixed output voltage.

Figure 6.8b is a photo of the converter and controller module. It shows the main controller, AC–DC rectifier, DC–AC inverter, and DC–DC converters inside. The back panel of this module is populated with AC and DC power ports and digital I/O ports (RS232, CAN, user switch, etc.).

6.2.2.3 Charge Transfer Interconnect Capacitor Module

The proposed HEES system has five EES nodes including three EES banks, AC input, and AC output. The maximum number of possible simultaneous independent charge transfers among the five EES nodes is only two. Therefore, we employ the shared-bus architecture to the HEES system for simplicity among several CTI architectures, which is shown in Fig. 2.2c.

The CTI is positioned in the middle of power converters as the input or output of them, and so a sufficient amount of bulk capacitors is required to secure voltage stability in order to cope with large transient current variations. The power converters have limited capability to deal with sudden changes of the load current, which may result in rapid changes of the CTI voltage unless the CTI has enough energy buffer. Maintaining a desired CTI voltage level is important because non-optimal CTI voltage leads to degraded power conversion efficiency of the power converters. What makes the problem worse is that the CTI voltage drop may violate the minimum input voltage requirement of power converters and result in system failure.

We implement the energy buffer for the CTI with an array of large-capacitance aluminum capacitors. Figure 6.8c is a photo of the CTI capacitor module. We connect six of United Chemi-Con 22,000 μF aluminum capacitors in parallel to compose the CTI capacitor array. The total 132,000 μF capacitance is proven by experiments to be sufficient to maintain a stable voltage against 300 W load increase when charged over 25 V. The main controller determines the CTI voltage level with considerations on the energy efficiency of the charge transfers and predicted load demand. The CTI capacitor module has one voltage meter and two current meters on the front panel to show the CTI voltage, current from CTI to the load devices, and current from the power grid to the CTI. The emergency switch is also placed on the front panel to cut off the main power to inactivate all the power paths manually in case of emergency.

6.2.2.4 Bidirectional Charger

The charger design is a key for the HEES system to achieve high efficiency and high reliability. It has integrated functionalities of a DC–DC converter, battery charger, and battery monitor. This is similar to conventional battery management systems, but provides more high-level functionalities for the HEES system management by being integrated with the main controller and other chargers. Specifically, the bidirectional charger implemented for the proposed HEES system has the following features:

- Bidirectional conversion: It is capable of charging the EES array from the CTI, as well as discharging the EES array to the CTI.
- Voltage and current regulation: It generates either a regulated voltage or regulated current. When it is charging the EES array, this is necessary to perform the constant-current and constant-voltage (CC-CV) charging. When it is discharging the EES array, this is necessary either for maintaining the CTI voltage or injecting designated amount of current into the CTI.
- Adjustable output regulation: The output voltage or current can be adjusted dynamically. The output voltage is determined by the EES array maximum voltage (when charging), or the desired CTI voltage (when discharging). The output current is determined by the main controller considering efficiency and reliability.
- Wide operational range: The terminal voltage of the EES arrays changes depending on their SoC, especially for supercapacitor array whose voltage is linearly proportional to the SoC. The CTI voltage also changes in a wide range for energy efficiency and reliability. Therefore, the charger has a wide input and output voltage range of 6–36 V.
- Board-to-board communication: It reports the current status of the bank to the main controller and receives operational commands from the main controller through the CAN bus.

Figure 6.9a shows the charger implemented for the proposed HEES system. The charger is composed of three parts: main converter (Ⓐ), power path controller (Ⓑ), and supercapacitor boot-up charger (Ⓒ). It is powered from a separated power rail, not from the CTI, for reliable operation of the system. Figure 6.10 shows the conversion efficiency of the main converter for different input and output voltage at 0.5 and 1.5 A load current.

The main converter performs unidirectional adjustable voltage or current regulation with wide input/output voltage (6–36 V) and high-current (up to 10 A). The converter circuit uses a combination of the LTC3789 DC–DC converter [3] and the LTC4000 power management IC [4] from Linear Technology. The LTC4000 is originally designed to generate a fixed voltage or fixed current for charging battery from the wall power. We modify the voltage and current feedback loops with digital potentiometers to make them dynamically adjustable from the microcontroller. We use the Texas Instruments LM3S2965 microcontroller to control the charger as for the main controller. The charger has a CAN bus interface to communicate with the main controller as well.

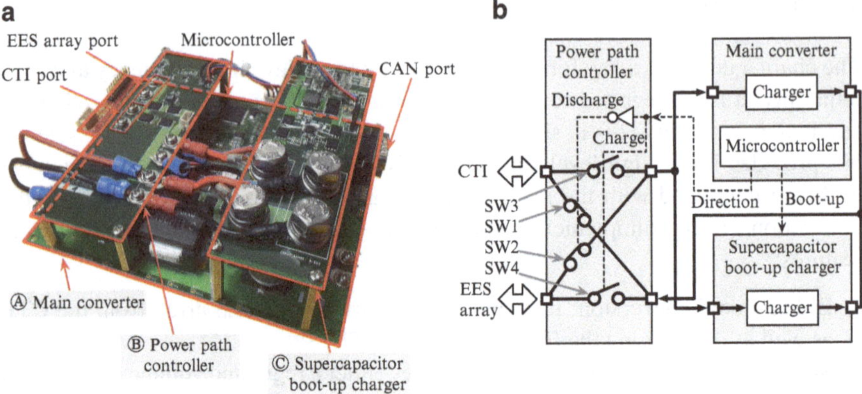

Fig. 6.9 Implemented charger and the power path of the charger. (**a**) Implemented charger. (**b**) Power path of the charger

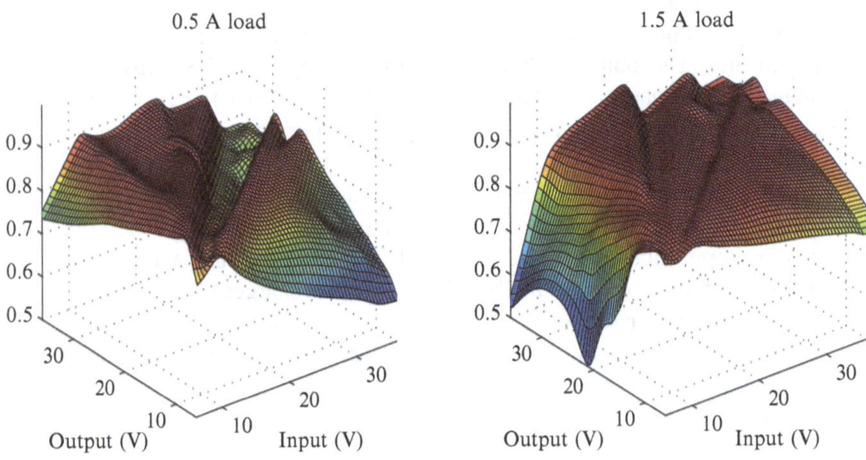

Fig. 6.10 Efficiency of the main converter

The power path controller swaps the input and output of the main converter as needed to enable bidirectional power conversion. When charging the EES bank, the input is the CTI and the output is the EES array; and when discharging the EES bank, it is the opposite. We use four solid-state relays made of a pair of MOSFET switches to dynamically change the input and output as shown in Fig. 6.9b. Two solid-state relays are coupled as a pair and opened or closed together, and the two pairs are exclusively closed. For example, the microcontroller closes SW1 and SW2 are closed and opens SW3 and SW4 as shown in Fig. 6.9b when charging. We implement a hardware protection circuit on the power path controller to prevent short circuit from the CTI and EES array by control flaws or signal glitches.

The supercapacitor boot-up charger is added to the charger for the supercapacitor bank module only. It is a fixed current buck mode switching charger made of the BQ24640 supercapacitor charger IC from Texas Instruments. The supercapacitor terminal voltage is linearly proportional to its SoC, different from the batteries that maintain relatively a constant terminal voltage regardless its SoC. Even though the main controller manages the terminal voltage of the supercapacitor array within the operational range of the main controller, sometimes we may have an under-voltage due to intended discharge (supercapacitor replacement) or unintended discharge (self-discharge). We use the boot-up charger to charge the supercapacitor array up to the minimum voltage level of 6 V that the main converter can handle in such cases.

The HEES systems has multiple chargers, one per each EES bank. We connect all the chargers through the CAN bus together with the main controller in order to perform systematic control of the HEES system. The chargers send voltage and current measurement results to the main controller through the communication network, and receive commands determined by the main controller based on the system management policies.

6.2.2.5 Supervising Control Software

The main controller and bidirectional chargers run μC-OS II RTOS to perform system management and EES bank management, respectively. The main controller mainly performs supervisory tasks for global HEES system management such as the charge management, while the chargers perform individual control of the EES bank. Figure 6.11 shows the block diagram of the control software implemented in the main controller and chargers.

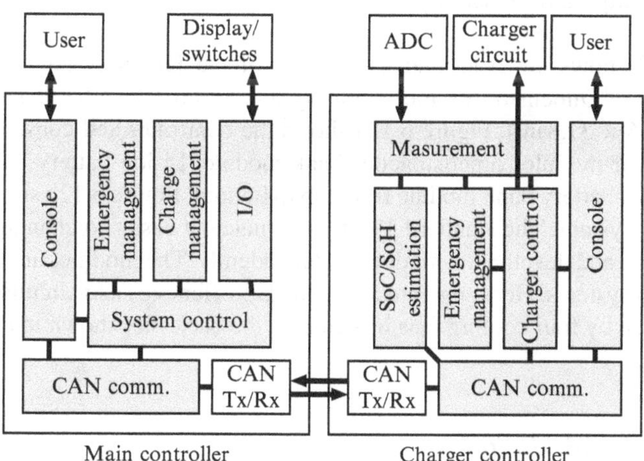

Fig. 6.11 Design of the main controller and charger controller. Lines between tasks denote inter-task communication

The main controller is in charge of determining the CTI voltage and current of each EES bank for high energy efficiency and reliability. We implement the charge management task in the main controller to determine the optimal values [9, 10, 12, 13]. The decision is made based on many parameters including voltage, current and SoC of the EES banks and load demand prediction. We do not cover the optimization methods minutely in this dissertation. The system control task makes the overall decision on energy flow of each EES node based on decisions of other tasks.

The charger is in charge of measuring the status of the EES bank and controlling the power conversion. The measurement task measures the voltage/current of the input/output of the power converter. The SoC/SoH estimation task calculates the SoC and SoH based on the measured voltage and current profiles. The converter control task changes the voltage/current regulation feedback loops according to the commands from the main controller.

There are also some common tasks implemented in both the main controller and chargers. The CAN communication tasks collect messages from other tasks and deliver them over the CAN network layer periodically or on demand. Critical messages related with the main controller's decision (e.g., measured voltage/current and control commands) are delivered periodically with a high priority; while slowly changing values (e.g., SoC and SoH) are delivered upon request with a low priority. The console task provides manual control interface to the user which overrides the automated control. The emergency management task has a highest priority to detect abnormal status and shut-down all power paths and power converters in the system. The console task is for communication with the user for monitoring and manual control.

6.2.2.6 Component Assembly

Figure 6.12 shows the front and back views of the HEES system assembled in a 19-in. rack. Dimension of the whole system is 60 (W) × 65 (L) × 96 (H) cm (23.6 × 25.6 × 37.8 in.). Figure 6.12a shows the controller and converter module, CTI capacitor module, supercapacitor bank module, Li-ion battery bank module, and lead-acid battery bank module from top to bottom. Figure 6.12b shows the back view of the system. The standard 19-in. rack makes it easier to change the system configuration and develop the modules independently. The modules are reusable for other HEES systems. Heat generated from the EES elements and circuits is removed from the rack by four cooling fans installed on the back (not shown in the figure).

6.2.3 Control Method

The CTI should maintain a certain voltage level in order to avoid under-voltage failure when the load current increases suddenly. Control of the CTI voltage is

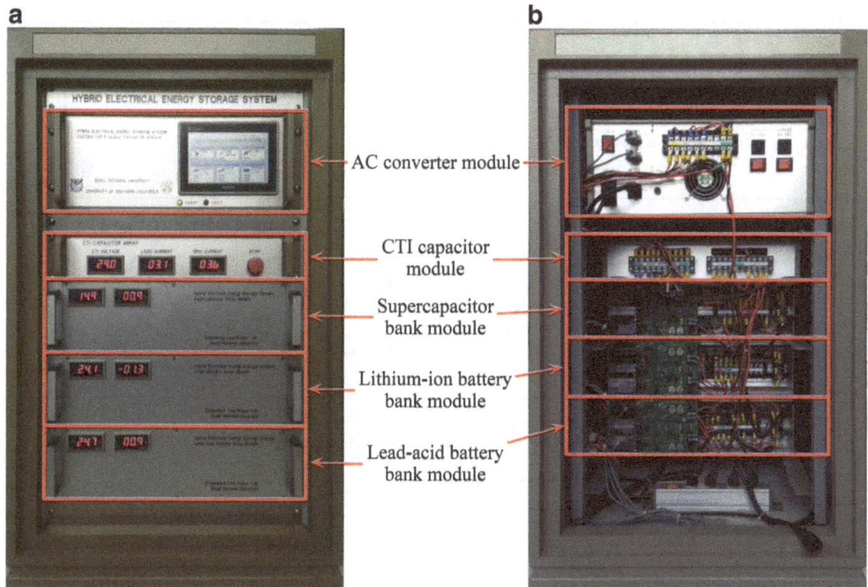

Fig. 6.12 Front and back views of the assembled HEES system. (**a**) Front. (**b**) Back

critical to energy-efficient and reliable operation of the HEES system. The net current, sum of inflow currents and outflow currents, of the CTI should be zero in steady state in order to keep the CTI voltage constant. In an ideal case, we may operate all power converters in the current-regulating mode and adjust the output current immediately responding to CTI voltage variation. However, the capacitance of the CTI capacitor is not very large in practice, and thus the voltage changes very rapidly even with a slight mismatch between inflow and outflow currents. The software control is not fast enough to detect the CTI voltage variation and re-determine the current of EES banks in time. This is especially critical when experiencing a sudden increase of the load current but the system fails to increase the discharging current in time, resulting in the under-voltage failure.

We employ a cascaded control loop to overcome the CTI voltage reliability problem. We operate one of the discharging power converters in the voltage-regulating mode so that it regulates the CTI voltage with the hardware feedback control loop. The voltage regulation performed by the hardware voltage feedback control loop of the converter is fast enough to keep the CTI voltage at the desired level against the fluctuation of the input and output current of the CTI. Other current-regulating power converters extract or inject a designated amount of current from or to the CTI. Current supplied by the voltage-regulating power converter is consequently determined to the amount that makes the net current of the CTI zero. Choosing the voltage-regulating converter is an important decision though any of chargers of discharging EES banks and power converters from power sources can take this role. It is related to reliability because the it may have to solely maintain

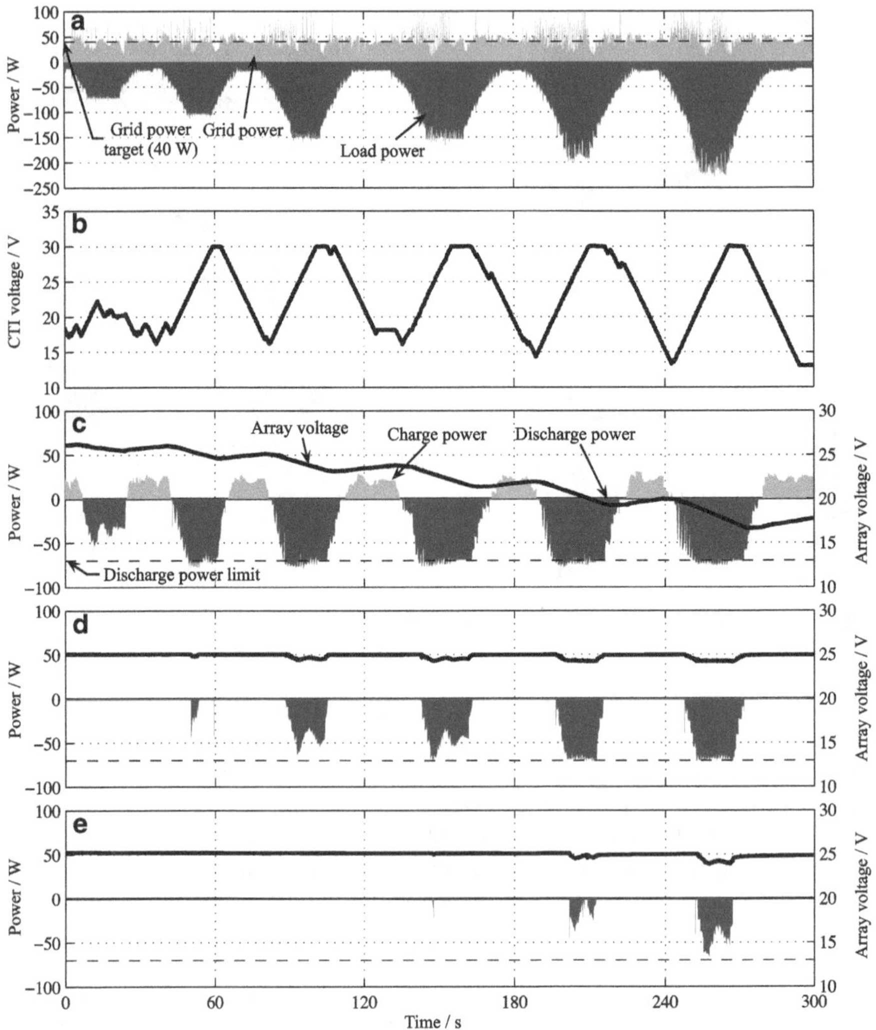

Fig. 6.13 Voltage and current measured during operation. (**a**) Load and grid power. (**b**) CTI voltage. (**c**) Supercapacitor. (**d**) Lithium-ion battery. (**e**) Lead-acid battery

the CTI voltage during transient periods. We designate the DC–DC converter from the power grid as the voltage-regulating converter for reliability because the power grid has virtually unlimited power capacity and energy capacity. Consequently, the bidirectional chargers in the EES banks are in current-regulating mode.

This is similar to conventional control method introduced for shared bus HEES systems at first glance [7, 8, 14]. However, the current control of each power source/EES bank is coupled with other power source/EES bank with a priori knowledge on them, and so limited in versatility and scalability. Our control method provides more flexibility and a higher degree of freedom for energy-efficient control by allowing arbitrary current for each EES banks as well as variable CTI voltage.

Figure 6.13 shows the results of the proposed control method. It shows voltage, current, and power measured from various points while the load power significantly changes over time. Figure 6.13a shows the load power (discharged from the HESS) and grid power (charged to the HESS). The dashed line is the grid power target, which is 40 W. The grid power stays near the target level even the load power changes between 15 and 220 W. As the load power increases, a higher CTI voltage becomes more beneficial to reduce conductance loss in the converter circuit. As a result, the CTI voltage changes in a wide range to achieve the best conversion efficiency as shown in Fig. 6.13b. Figure 6.13c–e show power and voltage variation of the supercapacitor array, Li-ion battery array, and lead-acid battery array, respectively. While the grid power is lower than the target, the supercapacitor bank is charged first. The array voltage of the supercapacitor bank increases or decreases as it is charged or discharged. The Li-ion battery bank and lead-acid battery bank consecutively start to discharge as the load power becomes heavier. This experimental result demonstrates the capability of operating high-level system management policies, which is enabled by the proposed control method.

References

1. Chakraborty S, Lukasiewycz M, Buckl C, Fahmy S, Chang N, Park S, Kim Y, Leteinturier P, Adlkofer H (2012) Embedded systems and software challenges in electric vehicles. In: Proceedings of the Design, Automation & Test in Europe Conference & Exhibition (DATE), pp 424–429
2. Kim Y, Koh J, Xie Q, Wang Y, Chang N, Pedram M (2014) A scalable and flexible hybrid energy storage system design and implementation. Journal of Power Sources 255(0):410–422
3. Linear Technology (2010) LTC3789: High efficiency, synchronous, 4-switch buck-boost controller
4. Linear Technology (2011) LTC4000: High voltage high current controller for battery charging and power management
5. Panasonic (2005) Value-regulated lead acid batteries: individual data sheet: LC-R123R4P
6. Samsung SDI (2009) Specification of product for lithium-ion rechargeable cell model ICR18650-26F
7. Thounthong P, Raël S, Davat B (2009) Energy management of fuel cell/battery/supercapacitor hybrid power source for vehicle applications. Journal of Power Sources 193(1):376–385
8. Thounthong P, Chunkag V, Sethakul P, Sikkabut S, Pierfederici S, Davat B (2011) Energy management of fuel cell/solar cell/supercapacitor hybrid power source. Journal of Power Sources 196(1):313–324
9. Wang Y, Kim Y, Xie Q, Chang N, Pedram M (2011) Charge migration efficiency optimization in hybrid electrical energy storage (hees) systems. In: Proceedings of the ACM/IEEE International Symposium on Low Power Electronics and Design (ISLPED), pp 103–108
10. Wang Y, Xie Q, Pedram M, Kim Y, Chang N, Poncino M (2012) Multiple-source and multiple-destination charge migration in hybrid electrical energy storage systems. In: Proceedings of the Design, Automation & Test in Europe Conference & Exhibition (DATE)
11. Wiehagen J, Harrell D (2001) Review of residential electrical energy use data. Tech. rep., NAHB Research Center, Inc.

12. Xie Q, Wang Y, Kim Y, Chang N, Pedram M (2011) Charge allocation for hybrid electrical energy storage systems. In: Proceedings of the IEEE/ACM/IFIP International Conference on Hardware/Software Codesign and System Synthesis (CODES+ISSS), pp 277–284
13. Xie Q, Wang Y, Kim Y, Shin D, Chang N, Pedram M (2012) Charge replacement in hybrid electrical energy storage systems. In: Proceedings of the Asia and South Pacific Design Automation Conference (ASP-DAC), pp 627–632
14. Zhang Y, Jiang Z, Yu X (2008) Control strategies for battery/supercapacitor hybrid energy storage systems. In: Proceedings of the IEEE Energy 2030 Conference, pp 1–6

Chapter 7
Conclusions and Future Directions

EES systems offer various benefits of improved energy efficiency, reliability, availability, and cost-effectiveness for wide range of applications including the power grid, renewable power sources, EV/HEV, and so on. It is a practical approach to implement EES systems with the current EES element technologies, where each of them has unique strengths and weaknesses. However, while design and control of the conventional homogeneous EES systems are relatively straightforward, the HEES systems requires elaborated design and control methods to maximize its benefits over homogeneous EES systems.

This book studied the design considerations for practical implementation and deployment of the HEES systems. We proposed high-level HEES system architecture design and control methods to satisfy the design considerations. The proposed architecture-level HEES designs, which are the reconfigurable EES bank architecture and networked CTI architecture, aim at reducing the maximizes the energy efficiency by reducing the energy loss induced by power conversion. We introduced optimization problems and systematic solution methods involved in the proposed architectures. In addition, we proposed a design and operating method called the MPTT that achieves the joint optimality of HEES system with renewable power sources. We also introduced our HEES system prototype with detailed practical discussion on each component design.

The following issues are remained for future research:

- We will devise a method to find the optimal sequence of charge transfers. This is an important step to schedule the operations in the HEES system from the power supply/demand profile. We will first try to find the optimal sequence of charge transfers assuming the HEES system is given. Ultimately, the optimal sequence of charge transfers should be simultaneously derived with the optimal HEES system design.

Y. Kim, N. Chang, *Design and Management of Energy-Efficient Hybrid Electrical Energy Storage Systems*, DOI 10.1007/978-3-319-07281-4_7

- Optimization of HEES systems only is not enough for the true optimal of energy system. The power generation and power consumption should be optimized simultaneously. Power sources, HEES systems, and load devices should be aware of each other not only in design time, but also in runtime, to achieve the holistic energy optimization.
- We will implement the charge management schemes in the HEES prototype and prove their practicality. This involves not only issues on theoretical optimality, but also practical issues such as reliable control and feasibility of real-time computation. We will refine the management schemes with considerations on those issues and devise an effective implementation.

Zeitfracht Medien GmbH
Ferdinand-Jühlke-Straße 7
99095 Erfurt, Deutschland
produktsicherheit@kolibri360.de